Europe at the Polls

EUROPE IN TRANSITION: THE NYU EUROPEAN STUDIES SERIES

The Marshall Plan: Fifty Years After
Edited by Martin Schain

Europe at the Polls: The European Elections of 1999
Edited by Pascal Perrineau, Gérard Grunberg, and Colette Ysmal

Europe at the Polls:
The European Elections
of 1999

Edited by

Pascal Perrineau, Gérard Grunberg, and Colette Ysmal

palgrave

EUROPE AT THE POLLS
© Pascal Perrineau, Gérard Grunberg, and Colette Ysmal, 2002

Softcover reprint of the hardcover 1st edition 2002 978-0-312-23895-7

First published 2002 by
PALGRAVE
175 Fifth Avenue, New York, N.Y. 10010 and
Houndmills, Basingstoke, Hampshire RG21 6XS.
Companies and representatives throughout the world

PALGRAVE is the new global publishing imprint of St. Martin's Press LLC Scholarly and Reference Division and Palgrave Publishers Ltd (formerly Macmillan Press Ltd).

ISBN 978-1-349-63226-8   ISBN 978-1-137-04441-9 (eBook)
DOI 10.1007/978-1-137-04441-9

Library of Congress Cataloging-in-Publication Data

Vote des Quinze. English.
    Europe at the polls: the European elections of 1999 / edited by
Pascal Perrineau,
Gérard Grunberg, and Colette Ysmal.
        p.cm. (Europe in transition, the NYU European studies series)
    Includes bibliographical references and index.
        1. European Parliament—Elections, 1999. 2. European Union countries—
Politics and government—1989. I. Perrineau, Pascal. II. Grunberg, Gérard. III.
Ysmal, Colette. IV. Title.

JN36 . V67513 2001
324.94'0559—dc21   2001053263

A catalogue record for this book is available
from the British Library.

Design by Newgen Imaging Systems (P) Ltd, Chennai, India.

First edition: January, 2002
10 9 8 7 6 5 4 3 2 1

Contents

Part One
Europe, Citizens, and Political Elites

Part Two
A Europe of Political Forces

Part Three
A Europe of Voters and Institutions

List of Figures and Tables

Figures

Tables

viii • List of Figures and Tables

Appendices

Contributors

BRUNO CAUTRÈS, Researcher, CNRS, Director, Information Centre for Socio-Political Data (CIDSP, Grenoble)

PASCAL DELWIT, Professor of Political Science, Director, Center for the Study of Political Life, Free University of Brussels (ULB)

LIEVEN DE WINTER, Professor of Political Science, Catholic University of Louvain and Catholic University of Brussels

ANDRÉ-PAUL FROGNIER, Professor of Political Science, Catholic University of Louvain

JACQUES GERSTLÉ, Professor of Political Science, University of Paris-Dauphine

GÉRARD GRUNBERG, Director of Research, CNRS, Centre for the Study of French Political Life (CEVIPOF, Paris)

DAVID HANLEY, Professor of European Studies, University of Cardiff

PIERO IGNAZI, Professor of Political Science, University of Calabre

MARC LAZAR, Professor History and Political Sociology, IEP of Paris

CHRISTOPHER LORD, Professor of Political Science, Jean Monnet Study of European Parliaments Chair, University of Leeds

PABLO MEDINA LOCKHART, Geography, Researcher, Free University of Brussels

GERASSIMOS MOSCHONAS, Professor of Political Science, Panteion University of Athens

FERDINAND MÜLLER-ROMMEL, Professor of Political Science, University of Lüneburg

PASCAL PERRINEAU, Professor of Political Science, IEP of Paris, Director of CEVIPOF (Paris)

HERMANN SCHMITT, Director of Research, Mannheim Center for European Social Research (MZES), University of Mannheim

KLAUS SCHOENBACH, Professor, Amsterdam School of Communication, University of Amsterdam

HOLLI A. SEMETKO, Professor, Amsterdam School of Communication, University of Amsterdam

RICHARD SINNOTT, Professor of Political Science, University of Dublin, Centre for Comparative Research on Public Opinion and Political Behaviour

JACQUES THOMASSEN, Professor of Political Science, University of Twente

CHRISTIAN VANDERMOTTEN, Professor of Geography, Free University of Brussels

MARINA VILLA, Professor of Theory and Information Systems, Catholic University of Milan–Brescia

COLETTE YSMAL, Director of Research FNSP, CEVIPOF (Paris)

Foreword

Martin A. Schain, Series Editor

Europe at the Polls is the second volume in the NYU-European Studies series, *Europe in Transition*. In this series we explore the core questions facing the new Europe. We regard this study as simply the best comprehensive book that has been published on the elections for the European Parliament. Twenty-two distinguished scholars from six European countries have joined forces to produce a study, not only of the elections themselves, but also of the relationship of these elections to the political structures and political forces in Europe.

The analyses of political parties across the 15 countries of the European Union in the second section provide a rich source for scholars and students who are interested in the comparative politics of Western Europe. If we read these chapters together with those in the first section, we get a sense of how little progress there has been in the development of trans-European political parties after five elections of the European Parliament by universal suffrage. However, we also learn that the "European dimension" has gained considerable autonomy in determining the left–right division, and that the European orientation of the representatives is considerably greater than that of the voters they represent. In addition, scholars who work on parties and elections have become increasingly European in their orientation. As we can see from this volume, leading scholars across the EU now routinely collaborate together.

The annexes at the end of this book ensure that this study will be a reference work for data on European politics for many years. They help to create a cross-national portrait of European parties and elections at a single point in time, and will thus serve as an important basis for future scholarship.

Introduction

Gérard Grunberg, Pascal Perrineau, and
Colette Ysmal

This book is different than the previous book we had written on the 1994 European elections: *Le vote des Douze* (Presses de Sciences Po., 1995). First, it is less focused on election results, electoral behavior, or ideological and political attitudes. One of the reasons for this is that, for those who are not concerned with reducing the European elections to their purely French dimension, such an analysis is extremely difficult to conduct in a short period of time when one must work on 15 countries with different traditions of both recording and disseminating electoral data. Furthermore, with the exception of Europe-wide barometers that do not deal with elections, a central election study center that operates during the voting does not exist.[1] We nevertheless felt it important to provide the reader with appendices in which we present all the data available. This book also intends to consider what these elections revealed about the construction of Europe. Therefore, we take a dual approach. The first approach sets out, whatever the subject considered—citizens, the elites, the election campaign, political forces, electorates, institutions—to broach the subject from an overall standpoint, highlighting national characteristics only to account for differences or problems that remain within the European Union (EU). The second calls upon French and European specialists on these issues who are working either independently, in pairs, or in groups.

If one refers to the 1999 elections, it must be acknowledged that the EU is simultaneously an edifice that is recognized and accepted by the majority of its citizens and subject to crisis and disputes. Public opinion is not unanimous on either the policies that the Union should implement, its institutions, or its future. With economic and monetary union barely achieved, some critics have pointed to the absence of a social policy. Others have challenged the primacy of the economy and finance and regretted that the construction of a united Europe concerned itself so little with the values upon which it was founded. To this scenario was added the malfunctioning of the Commission and the beginning of a power shift in favor of the Parliament. Finally, it must be kept in mind that the question of expanding the EU is bound to challenge the very principles of its functioning, a subject on which the European elites are so vague that voters can barely grasp the stakes. Once again, as in past elections, Europe was difficult to interpret and therefore has little chance of structuring voter choices. And yet, with each successive European election, it indeed seems that a European question is taking shape, as is a cleavage that has gained its autonomy from the traditional left–right divide.

Bruno Cautrès and Richard Sinnott demonstrate in chapter 1 how the rationales that structure the attitudes of the citizens of the Europe of 15 nations with respect to the EU are more rooted in the diversity of social and cultural positions and in the heterogeneity of national cultures than in political positions along the left–right axis. Having emerged as a concrete issue at the economic and political levels, Europe has moved away from the permissive consensus from which it once benefited to become an autonomous issue that does not easily overlap categories of left and right. One could have imagined that, escaping in this manner from the standard categories of political debate, the European issue would foster a major crisis of political representation. Hermann Schmitt and Jacques Thomassen (chapter 2) show, on the contrary, that political representation is working better than commonly believed at the European level. Of course, when one looks at the concrete European policies that are carried out, a large gap exists between the voters and their representatives, the latter being more European than the former. On the other hand, at the level of the general orientations of European construction, the connection between the people and those who represent them is close and entirely satisfactory. The two authors also note that since election of the European Parliament by direct universal suffrage was instituted in 1979, political representation in Europe has changed considerably. The left has become more European, whereas the right has shifted from Europhilia to the most blatant form of Euroskepticism. But it is especially the European dimension (increased integration versus the reaffirmation of national independence) that has gained an increased autonomy with regard to the left–right dimension. The European dimension affected the electoral body and brought about an evolution among the elites, leading the authors to conclude that a "representation from below" has imposed itself at the European level, with voters initiating change within the elites more than the elites influencing voters. Because of the autonomy of the European issue, European elections can no longer be apprehended using the model of a "second-order election." The European elections are no longer, as André-Paul Frognier observes in chapter 3, "meaningless," nor are they elections whose logic comes from national votes, but rather they must be considered as elections with their proper political logic. André-Paul Frognier demonstrates this by showing how voter participation in these elections was more structured by the level of European identification of the various national electorates than by the participation rate of these same electorates in the last national elections.

Jacques Gerstlé, Holli Semetko, Klaus Schoenbach, and Marina Villa (chapter 4) show, moreover, in their comparative study of election campaigns in France, the Netherlands, Germany, and Italy, that the flagging Europeanization of these campaigns is not due to a rejection of the European Union or to an internal instrumentalization of these campaigns but to the fact that the June 1999 European elections were blurred by a major international event (the war in Kosovo) and sometimes by a very heavy national agenda (Germany). Given the growing independence of the European issue among public opinion and European citizens, one can have the impression that the system of political forces remains somewhat behind.

Two points of view can be taken with regard to the political forces. The first has to do with the difficult conditions of establishing the European Union and to the slowness of creating truly European political groups. The European People's Party (EPP) has certainly gained strength and increased its role in the reorganization of the moderate right, as David Hanley and Colette Ysmal note in chapter 9. At the same time, the Party of European Socialists (PES) has, as is demonstrated by Gérard Grunberg and Gerassimos Moschonas (chapter 6), developed since 1992 and set up activities at the level of the 15 countries (think-tanks and programs). Nevertheless, these large parties, masters of the political game at the European Parliamentary level, have a major problem in imposing themselves upon national public opinions, among whom they are in fact little known and who do not make up their minds according to the parties' positions. The failure frequently comes from the parties' inability to present a program and to put forth something other than a catalogue of proposals from which any potentially conflictual aspect is eliminated.

What is true of the two largest European parties is even more so of the smaller parties. The environmentalists organized in the European Parliament in the group GGEP (Greens Group in the European Parliament) have indeed published a series of common proposals, but these have not been taken up by the corresponding parties at the national level (Ferdinand Müller-Rommel in chapter 7). In the same way, the ethnic-regional parties described by Lieven de Winter in chapter 8 have a common structure—the Democratic Party of the Peoples of Europe–Free European Alliance (PPDE-ALE)—but this group has no integrative capacity whatsoever. Finally, the Communist and extreme-left galaxy (Marc Lazar in chapter 5) finds itself totally incapable, this time for political reasons, of constituting anything else than a technical entity, that is, ensuring the existence of a Communist group in the Parliament. The divisions are indeed too strong both among the various heirs of Communism and between these and the descendants of Trotskyism or Maoism. The same is currently true for the extreme right following the scission of the National Front in France and the failure of the two parties that emerged from it; the transformation of the MSI into Alleanza Nazionale; and the reticence of the latter, as well as that of the Liberal Party in Austria (FPÖ), to commit itself to the extreme right (Piero Ignazi and Pascal Perrineau in chapter 10).

The second method of examining the political forces is to concentrate on results. The 1999 elections have been interpreted as a victory for the EPP and a stinging defeat for the PES. This is incontestably true, but it is necessary to analyze the situation at closer range. The two chapters by David Hanley and Colette Ysmal and by Gérard Grunberg and Gerassimos Moschonas are complementary: the first questioning the illusion of success on the part of the EPP, and the second dealing with "an honorable defeat on the part of the Socialists." For both parties, in fact, the balance of power was essentially inverted in two countries—Great Britain and Germany—while, in the other countries, the evolution of the vote was generally moderate and never to the exclusive advantage or detriment of one force or the other.

The most significant changes once again affected the most marginal parties. As Marc Lazar demonstrates, the parties that emerged from Communism remain in a difficult electoral position and are now in competition with a reinvigorated extreme left. The environmentalists obtained an indisputable success resulting, according to Ferdinand Müller-Rommel, from the rallying of young European voters stemming from the middle classes and sensitive to the themes of postmaterialism and the new politics. The ethnic-regional parties often made a respectable showing, according to Lieven de Winter, due to the advantages that these parties can draw from European integration. The principal victim of the European elections was the extreme right, which, following the successes of the 1980s and the beginning of the 1990s, was reduced to a bare minimum (Piero Ignazi and Pascal Perrineau).

An analysis of the families of political parties leads to a study of the electoral geography carried out by Christian Vandermotten and Pablo Medina Lockhart (chapter 11). Working with the results of the last European elections in 491 regions of the 15 countries of the European Union, they established a precise typology of European political families, distinguishing nine groups divided into three large families (left, center and classical right, extreme right). We discover a leftist Europe frequently rooted in peripheral zones where there is opposition toward the central state in the old centers of heavy industry and in the central metropolitan zones. The right is more present in a mid-European space that runs from Great Britain to northern Italy, passing through the Rhine region within a pericentral space that maintains a marked rural character and in certain peripheral zones marked by religious tradition or the weight of small rural agriculture. Whether they are on the left or right, these voters have diverse attitudes toward the Parliament in Strasbourg.

The final two chapters of this book deal with the central issue of the relationship between the European citizen and the European Parliament. Do the European elections legitimize the European Parliament, thereby legitimizing the other European institutions? Furthermore, are these elections their only source of legitimacy? The two contributions by Pascal Delwit (chapter 12) and Christopher Lord (chapter 13) focus on this question and reply to each other indirectly.

Pascal Delwit, who studies the evolution of voter participation in the European elections, claims that the continuing drop in the voting level since 1979—from 55 to 48 percent between 1994 and 1999 in the countries of the Union where voting is not obligatory—weakens the legitimacy of the European Parliament. He believes that this drop in participation is the result of a situation in which the voter does not have the possibility of expressing himself on the overall problems of the future of the Union. The differences between the various political families are not apparent, the stakes are not clear, and a de facto majority is nonexistent. The voter cannot, as in classical parliamentary systems, promote or sanction specific policies and political teams.

Delwit also questions the negative effects of a noncampaign, with the European elections being nothing more than the juxtaposition of numerous national campaigns whose stakes are weak and purely internal. All of this leads to a declining legitimacy of the European Parliament and the persistence of a democratic malaise. For him, attempts to make European elections legitimize European institutions through universal suffrage are a failure.

Christopher Lord, based on a study of the political dynamics of the European Parliament, presents a largely different viewpoint from the preceding one. Taking the opposite view with regard to the theory that the European elections are of a secondary order or are essentially national by-elections, he believes them to be of major importance in the formation of each European Parliament. There exists an interaction among the elections that take place every five years, the emergence of a model of European Parliamentary politics, and the manner in which the new Parliaments model their own institutional development. According to Lord, before concluding too quickly that there is a contradiction between the increasing powers of the European Parliament and the decline of its electoral legitimacy, the theory of "second-order elections" should be regarded with great circumspection.

The other element of his argument is that the European Parliament's legitimacy is not based only on the elections in a system in which the member states have their own legitimacy and in which the application of a classic parliamentary model, which should increase voter participation, would in return result in delegitimizing effects in a multipolar (multistate) political system. Finally, legitimacy can also come, under these conditions, from the capacity of the Parliament to impose upon itself a certain restraint in the exercise of its powers. Thus Lord affirms that the real test for the European Parliament is measured in the ability to establish an optimal balance among voter mobilization, the diffusion of power, and consensual methods.

Concerning the European political system, he believes that there is indeed a system of European political parties and that the European elections play a central role in the manner in which this system is constituted and evolves. The victory of the EPP and its alliance with the Liberals in order to elect the new president in 1999 does not necessarily mean that the practice of large EPP–PES coalitions is going to give way to the establishment of a lasting left–right cleavage, even if the latter is important, but could on the contrary restructure the relationship between the two major parties as well as their methods of cooperation. In conclusion, according to Christopher Lord, the challenge for the recently elected Parliament is to establish a mixture in which competition and consensus, in both the relationship with the Commission as well as the relationship within the different political tendencies, would be applied in the different moments of the Parliamentary process.

The two authors, despite their different perspectives, agree on the weakness of the existing relationship between the Parliament and the citizen. As Lord recognizes, even if there exists a system of European parties, "its electoral link capacity for representation relies too heavily on a happy co-incidence between the left–right character of the MEPs' agenda and the principal dimensions of national politics where MEPs are elected, rather than on any direct contribution by the Parliament to the development of a public forum on Union affairs."

Translated from the French by Eduardo Cué.

Notes

1. A survey was conducted by the *Political Representation in Europe* group. However, besides the fact that the results are understandably not immediately available to the entire scientific community, they were not available to the group's members before this book was written.

PART ONE

Europe, Citizens, and Political Elites

CHAPTER 1

The 1999 European Parliament Elections and the Political Culture of European Integration

Bruno Cautrès and Richard Sinnott

Introduction

European elections take place against the background of the political culture of European integration. "Political culture" in this context means the attitudes, preferences, perceptions, knowledge, and behavioral propensities of the mass of European citizens vis-à-vis the European institutions in particular and European integration in general. In order to investigate the culture of integration in which the European Parliament elections of 1999 took place, this chapter first provides an overview of the development of orientations toward European integration since the early 1970s, including some examination of the limited evidence that exists on the cognitive aspect of these orientations. The chapter then focuses on the situation just prior to the 1999 elections and summarizes the main features of attitudes toward Europe and its institutions as the citizens prepared (or, as we know, in many cases did not prepare) to go to the polls to elect the members of the first European Parliament for the new millennium. This section of the chapter also provides a more detailed examination of two indicators of attitudes to integration, analyzing variations across countries and across the different social and demographic sectors of European society. The third main section of the chapter compares the attitudes and perceptions of citizens as manifested in the 1994 and 1999 European elections. The comparison is made on the basis of postelection data collected by the Eurobarometer in June–July 1994 and in autumn 1999.

The fact that these two surveys used identical questions regarding the Parliament and the elections to it enables one to make direct comparisons between the situation in 1994 and in 1999. In particular, it enables one to examine, at least in a preliminary fashion, the effect that the resignation of the Commission in March 1999 and the events surrounding it had on attitudes to the Commission and the Parliament. The final section of the chapter seeks to pull these various strands of evidence together and to draw some general conclusions about the role of the European public in the current system of political representation at European level.

An Overview of Orientations toward European Integration

The measurement of support for European integration is highly sensitive to variations in question wording. The four standard Eurobarometer indicators of attitudes to integration are customarily labeled "unification," "membership," "benefits," and "dissolution." The levels of support for integration elicited by these questions depend both on the stimulus presented in the question and on the response categories, in particular on whether the response categories offer an explicit middle position.

The *unification* indicator measures support for a very general aspiration ("efforts to unify western Europe") and does so on a four-point scale ("very much for," "to some extent for," "to some extent against," "very much against") that does not provide an explicit middle or neutral point. The rather vague stimulus and the absence of a middle position combine to produce high levels of support for integration, touching almost 80 percent at the beginning of the 1990s (see figure 1.1). The *membership* indicator provides a more concrete stimulus (country X's membership in the European Union) and a three-point scale ("a good thing," "neither good nor bad," "a bad thing") that, as indicated, includes an explicit middle position. It therefore probably provides a more realistic gauge of support for integration, which typically runs some 10 to 20 percentage points behind the unification indicator. The third Eurobarometer indicator asks whether the respondent's country has *benefited* from membership in the Union, but, like the unification indicator, it does not provide a middle position (the response categories are "benefited" and "did not benefit"). Because it involves an element of perception as well as evaluation, it is not surprising that the benefits indicator registers a positive response that is slightly lower than that registered by the membership indicator. Indeed, the gap between the two might be greater were it not for the presence of a neutral category in the membership indicator and the absence of such a category in the benefits indicator. Finally, there is the *dissolution* indicator. This question poses the hypothetical situation of

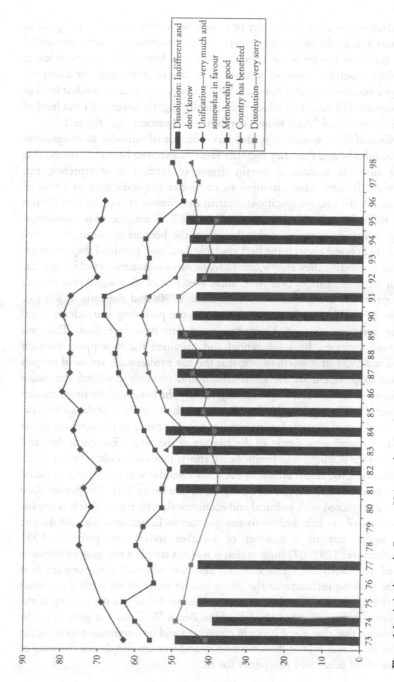

Figure 1.1 Attitudes to the European Union on four Eurobarometer indicators (unification, membership, benefits and dissolution), 1973–1998

the scrapping of the Community or Union, with strong negative and positive options and a middle position ("very sorry," "indifferent," "very relieved").[1] The indicator has been criticized because it is hypothetical; nonetheless it provides a useful measure of enthusiasm or lack of enthusiasm for European integration, showing, in June–July 1994 for example, a quite modest level of enthusiasm (43 percent), which was actually slightly lower than the level of "indifference" and "don't know" combined (46 percent) (see figure 1.1).

None of these questions is an ideal indicator of attitudes to integration; nor could it be said that they together form an adequate battery of items. One must therefore exercise a certain degree of caution in interpreting this evidence. In particular, attempts to categorize the indicators in terms of diffuse-affective versus specific-utilitarian dimensions of support quickly run into difficulties. As Niedermayer (1995: 54–5) notes, the only unambiguously utilitarian measure is that based on the benefits question. However, these are in many respects the best available data, and, provided they are interpreted cautiously, they enable one to make some reasonably valid inferences.

In terms of changes over time, there was, first of all, a significant decline in support for integration between the late 1970s and the early 1980s (see figure 1.1). This may have been related to the prevailing Eurosclerosis that many commentators on European integration have identified. This was followed, however, by a substantial and sustained rise in support between 1982 and 1991. It is worth noting that this rise predated the arrival of Jacques Delors as president of the Commission and certainly predated the major initiative of the first Delors presidency, namely the launch of the single market "Project 1992" program. On the other hand, there can be little doubt that the sustained rise in support for integration was due partly to the activism of the Delors Commission, partly to the passage of the Single European Act, and partly to the publicity and promotional efforts that surrounded Project 1992.

However, the actual arrival of the 1992 calendar year confirmed a general downward trend in support for integration (see figure 1.1). The period since 1989 is so packed with political and economic developments that it is impossible to attribute this decline to any one factor. Indications that the decline was under way in a number of member states even prior to 1991 (Niedermayer 1995: 67) suggest that it was not simply a response to the signing of the Maastricht Treaty but may also have reflected a negative reaction to the growing intrusion of the single market program on both the politics and the economics of individual states. Another factor was the waning of the euphoria that surrounded the fall of the Berlin Wall and the growth in the realization that that event brought challenges and uncertainties regarding the future shape and role of the Community and was not simply the dawning of a new era of peace and prosperity for all.

Examining the overall picture portrayed in figure 1.1, one can conclude that there is still fairly widespread support for the rather vague notion of "efforts to unify western Europe." However, as a result of a downturn that began in the second half of 1991, specific support for membership in the Union is not much above 50 percent. Moreover, when account is taken of the "dissolution" indicator, it becomes apparent that support for membership is matched by an almost equal level of indifference as to whether or not the Union continues to exist (see the bar graph in figure 1.1). All of this suggests that the famous "permissive consensus" (Lindberg and Scheingold, 1970), if it ever existed, was a rather fragile creature. One could go further and suggest that the term itself, so much bandied about, was actually misleading in that it glossed over significant flaws in the fabric of public opinion toward integration. Were it not for the prevalence of the permissive consensus assumption, there might not have been such surprise when, as the integration process began to make greater inroads into the economic and political life of the member states, support for integration began to wane, a waning that became manifest not only in opinion polls but also in referendums and in parliamentary debates in several countries.

Since detailed analyses of trends in support for integration throughout the European countries have been done in several other publications, we would like to focus on support for integration at the time of the 1999 European elections. Although the four indicators mentioned above have been measured regularly by Eurobarometer surveys, only two of them are available in the last study conducted before the 1999 elections (Eurobarometer 51.0, April–May 1999). We can use the technique of correspondence analysis to look at the main attitudinal dimensions captured by these two indicators and to plot the European countries as points in this two-dimensional space.

Attitudes to Europe in the Run-up to the 1999 EP Elections

Figure 1.2 shows the results of this correspondence analysis and illustrates the latent dimensions of attitudes toward Europe in European countries considered as a whole. The graphical display maps the first two dimensions of the multidimensional analysis, these being the dimensions that are the most significant in terms of the structure of attitudes. The first dimension, which is the most significant one (42.5 percent of the total inertia), represents the stronger opposition between pro- and anti-European opinions: On the right side of this horizontal dimension one finds the "has benefited" and "membership is good" categories and on the left side the opposite opinions. The second dimension (the vertical axis, 33.6 percent of the total inertia) pinpoints the opposition between nonfavorable opinions toward Europe and "don't knows." This dimension brings out the importance of the distinction

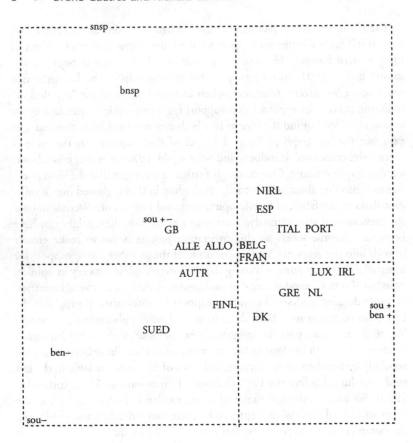

Figure 1.2 Correspondence Analysis of European Attitudes (Membership and Benefit Indicators; Countries as Supplementary Points. Eurobarometer 51.0, April–May 1999, 16177 Respondents.)

Source: Eurobaromètre 51.0 (March–April 1999), 16,177 simultaneously treated responses.
Sou−: EU membership is a bad thing; Sou+: EU membership is a good thing; Sou+−: EU membership is neither a good nor a bad thing; Snsp: undecided as to whether EU membership is a good or a bad thing; Ben+: the country benefited from EU membership; Ben−: the country did not benefit from EU membership; Bnsp: undecided as to whether the country has benefited from EU membership.
NIRL: Northern Ireland; ESP: Spain; ITAL: Italy; PORT: Portugal; GR: Great Britain; ALLE: East Germany; ALLO: West Germany; BELG: Belgium; FRAN: France; AUTR: Austria; FIN: Finland; SUED: Sweden; DK: Denmark; GRE: Greece; NL: the Netherlands; LUX: Luxembourg; IRL: Ireland.

between, on the one hand, indifference or absence of an attitude to integration and, on the other, actual opposition to integration.

As well as identifying the dimensionality of attitudes to European integration, the correspondence analysis shows that the location of the countries in the two-dimensional space is remarkably stable. The blocks of countries present a very familiar pattern: Support for European Union membership continues to be highest in Ireland, Luxembourg, and the Netherlands. It is also strong in Portugal, Greece, and Italy. As usual, this support is lowest in the U.K., former East Germany, and the new members. Sweden, Austria, Finland, France, and Belgium have an average position: They are close to the European average for support of integration and are, in any event, not strong supporters (for comparisons with previous analyses, see Cautrès 1998 and Cautrès 2000).

Preferences Regarding European Policy Integration

The correspondence analysis displayed in figure 1.2 gives an indication of the structure of attitudes toward Europe and illustrates the relative positions of the countries in this attitudinal space. The results are interesting in terms of general support for integration, but one must also look carefully and in detail at the policy content of support for Europe. If people do not, on average, strongly support the process of European integration, and if there is quite a lot of national variation in that regard, are there some domains of public policy in which people are ready to accept more European decisions, that is, in which they would like to see decisions taken at the EU level rather than at the national level? The Eurobarometer regularly measures the preferences of Europeans in terms of the level of decision making in certain areas of policy: Should they be decided by the national governments or jointly within the European Union institutions? As measured some weeks before the 1999 European elections, the average level of support for European decision making has been 52 percent, whatever the policy domain concerned. This average level of support for more European policymaking varies significantly according to the different domains of public action, as can be seen in table 1.1 (in which these domains are ranked according to the support for European-level decision making).

The first ten domains, showing the highest level of support for European policymaking, have a strong inherent transnational dimension: This is especially the case for foreign policy of the EU, the fight against drugs, humanitarian aid, research and technology, protection of the environment, and currency. For these policy areas, public opinion is largely in favor of the principle of more European decision making (the proportion varies between 60 and 70 percent). Other domains, for example immigration policy and rules for political asylum, also show a certain level of support for a European level of decision making, but this support is not very strong. It seems likely that public opinion considers these two domains as having both transnational and national dimensions.

Table 1.1 Preferred level of decision making by policy domain

Areas of policy should be decided ...	... jointly within the EU	... by national government	Don't know
Information about the EU	69	23	8
Foreign policy toward countries outside the EU	68	23	9
Fight against drugs	66	30	4
Humanitarian aid	65	29	6
Scientific and technological research	63	30	7
Currency	61	33	6
Fight against poverty and social exclusion	60	35	5
Support for regions having economic difficulties	59	33	8
Protection of the environment	55	40	5
Immigration policy	54	39	7
Rules for political asylum	52	40	8
Fight against unemployment	48	47	5
Defense	46	48	6
Agriculture and fishing policy	45	47	8
Rules for broadcasting and press	34	59	7
Cultural policy	32	59	9
Health and social welfare	30	65	5
Education	29	66	5

Source: Eurobarometer 51.0, April–May 1999.

In the economic policy domains there is a notable difference between, on the one hand, aid to disadvantaged regions and currency policy, two areas for which support for more European decision making is quite strong, and, on the other hand, the fight against unemployment, which elicits considerably less support for European-level decision making. The fight against unemployment is a domain that has national as much as European implications as far as public opinion is concerned. People are cautious about domains that touch on issues of welfare and are concerned about the consequences of European integration for national policies. Other domains, for example concerning sociopolitical values (media and cultural policies, education) or concerning sensitive issues like agriculture and fishery policies, also show that, for European public opinion, there are limits to support for more European-driven public action. These limits are quite stable through time.

Our analysis underlines the existence of conflict between two different models: on one side the "nation-state model," still there regarding issues like defense, welfare, and public benefits (unemployment, health, and social security), and preservation of national cultural identities (rules for broadcasting and press, cultural policy, education); on the other side a "community model" largely accepted for policy areas having a strong transnational dimension. The existence of this opinion cleavage illustrates the fears of losing, through the process of European integration, the national cultural identities and the benefits and welfare outputs of the nation-state model (see Pierre Bréchon, Bruno Cautrès, and Bernard Denni, 1995).

One must be careful in interpreting this opinion cleavage. The existence of this cleavage in European public opinion does not mean that attitudes toward Europe are structured by attitudes toward the level of policy making. The analysis shows that, for European public opinion, there are limits to the Europeanization of the public policymaking process. But these data do not give any indication regarding the quality of these opinions. Are they informed? Are they consistent?

The high level of indifference to European integration indicated in figure 1.1 reinforces the point that one should examine the quality of attitudes to integration and, in particular, that one should test the hypothesis that the responses to questions about Europe may contain a significant proportion of what Converse (1964) referred to as "nonattitudes." This hypothesis is confirmed when one examines the evidence on preferences regarding the attribution of policy competence to the European level or to the national level in greater detail. The Eurobarometer has asked a standard question on this matter over many years (see Sinnott, 1995). The question has consistently elicited a very low nonresponse rate (an average of about 8 percent across different issues and different years). However, an exploratory question on the overall issue of policy attribution between the national and European levels, which was inserted in Eurobarometer 41.1, strongly suggests that nonattitudes to this issue may be quite widespread. The new question included the specific response category "I haven't really thought about it." This response was chosen by 26 percent of respondents, a proportion that, when combined with the 10 percent who spontaneously offered a "don't know" response, yields more than one-third of the sample who acknowledge that they do not have any opinion on the basic issue of the appropriate scope of decision-making competence of the European and national authorities (Blondel, Sinnott, and Svensson, 1998: 65–72). Further evidence from the same study suggests that many of those who were willing to express an opinion may not have had a very explicit or well thought out basis for their view. Those who did give a response to the question of the overall range of issues decided on by the European Union were asked, "When you say [insert response to previous question], is this a general feeling that you have about the European Union (European Community), or have you specific issues in mind?" This retrospective probe showed that only 17 percent of the sample had specific issues in mind in responding to the original question. This fact strengthens the view that attitudes to European integration may not be very well formed. These weaknesses in attitudes to integration are confirmed by the evidence relating to knowledge of the institutions of the Union. For example, perceptions of the actual decision-making competences of the Union seem to be highly inaccurate. This is shown by Eurobarometer data from 1995 on perceptions of the actual allocation of decision-making power between national governments and "the European Union level" over a wide range of issues. *Inter alia,* the data show that 38 percent of the European public believe that foreign policy is decided at EU level, and precisely the same proportion believe that defense matters are decided at EU level (Eurobarometer, 1995: B65). On any reading of the common foreign and security policy as of 1995, these perceptions were wildly inaccurate. The proportion seeing foreign policy as being "at least to some extent decided at the European Union level" should have been much higher, and the proportion seeing the same for defense should have been much lower. Lest it be assumed that people are getting it wrong in relation to the common foreign and security policy simply because this is an inherently difficult and remote area and one that is complicated by the existence of NATO, one should also note that the proportion of Europeans who perceive agricultural policy as "at least to some extent decided at the European Union level" is only 40 percent! More generally, research using a variety of indicators of knowledge of European affairs shows that such knowledge is low and that low levels of knowledge adversely affect the structure and coherence of attitudes to integration (Sinnott, 2000). In short, the evidence on the cognitive aspect of the orientations of citizens to the European Union confirms the view that well-structured, well-informed, and supportive attitudes commensurate with the current stage of integration have not in fact developed.

Sociopolitical Logics of Attitudes Toward Europe

The different analyses presented so far show that public opinion toward Europe varies through time and across countries. It also shows that the measurement of public opinion is sensitive to the level of measurement (general support or more precise opinions) and to the type of indicators used (support, knowledge, aspirations). But one important question is not answered by the above analysis: What are the social and political determinants of European attitudes? Are they the same in all European countries? Rather than analyzing these questions country by country, one can focus on a few countries representing the two main groups: the more pro- and the more anti-integration. Tables 1.2 and 1.3 give the distribution of the two indicators used in figure 1.2 (membership and benefit) by sociodemographics for four countries: Sweden and Britain for the countries that are less favorable to integration; the Netherlands and Portugal for the countries that are more in favor. The main implications of the data in these two tables are clear: Educated people, people having high-level occupations, and men are more in favor of European integration and give more positive assessments of their country's membership in the European Union. The percentage differences are in some cases stronger between occupations (especially between manual workers and professional/higher managers) than between levels of education. Contextual effects are also evident in tables 1.2 and 1.3: Dutch or Portuguese manual workers are less in favor of integration than are Dutch or Portuguese managers but are much more in favor of it than are Swedish or British manual workers.

Table 1.2 Support for membership in the European Union ("a good thing")

	Sweden	Great Britain	The Netherlands	Portugal	Europe (15)
Occupation					
Farmers	45	33	44	51	47
Shopkeepers and artisans	36	21	70	60	53
Professional and higher managers	62	56	88	79	67
Lower managers	43	43	74	79	67
Routine nonmanual employees	32	29	72	65	53
Manual workers	20	24	66	54	43
Age completed education					
15 years or less	23	25	53	51	42
16–19 years	29	27	74	65	49
20 years and more	44	57	79	81	63
Still studying	35	36	79	76	59
Gender					
Men	42	37	76	66	55
Women	27	25	69	53	47
Age					
15–24 years	34	31	81	73	55
25–34 years	30	27	77	73	53
35–44 years	34	34	74	65	53
45–54 years	33	37	72	49	53
55–64 years	35	27	66	58	50
65 years and more	37	27	73	36	44

Source: Eurobarometer 51.0, March–April 1999 (row percentages).

Table 1.3 Perceived benefit of membership in the European Union ("benefited")

	Sweden	Great Britain	The Netherlands	Portugal	Europe (15)
Occupation					
Farmers	47	33	75	62	54
Shopkeepers and artisans	23	29	73	73	55
Professional and higher managers	40	55	78	75	63
Lower managers	21	44	66	82	54
Routine nonmanual employees	19	26	68	75	53
Manual workers	13	24	55	63	43
Age completed education					
15 years or less	17	26	48	63	43
16–19 years	11	27	68	77	47
20 years and more	28	55	73	87	60
Still studying	25	36	78	74	55
Gender					
Men	26	39	76	66	54
Women	16	24	62	53	45
Age					
15–24 years	24	30	77	74	52
25–34 years	22	28	74	83	52
35–44 years	18	32	69	83	50
45–54 years	22	34	69	64	53
55–64 years	19	29	57	62	48
65 years and more	21	32	52	55	43

Source: Eurobarometer 51.0, March–April 1999 (row percentages).

The percentage differences between countries are often as large as the percentage differences within a country. The levels of support for European integration are strongly affected by country-specific contextual effects in every social group, as can be seen by comparing across the rows of tables 1.2 and 1.3.

Attitudes toward Europe are also affected by political factors, as shown in table 1.4. Among the politicization indicators in table 1.4, the leadership opinion index has the strongest impact and gives the largest variations in support for integration among the four countries. Media exposure also gives rise to variations but with a weaker impact. Left-right scale positions have a small impact, showing that European and left-right cleavages are relatively independent of each other. Indeed, in most countries the electorate is crosscut by the European cleavage, but it should be noted that this relative independence varies according to country.

Table 1.4   Support for Membership in the European Union ("a good thing")

	Sweden	Great Britain	The Netherlands	Portugal	Europe (15)
Left-right scale positions (recoded)					
Left	19	42	77	56	51
Center	29	32	74	65	52
Right	57	29	74	60	57
Exposure to media					
+ +	38	37	75	70	55
+	31	30	75	66	51
−	26	23	67	49	46
− −	29	22	70	39	38
Opinion leadership index					
+ +	48	51	76	60	62
+	39	44	79	70	58
−	26	29	72	56	49
− −	29	17	53	45	39

Source: Eurobarometer 51.0, March–April 1999 (row percentages).

The European Electorate in 1994 and 1999

Whereas the 1994 election was just another routine stage in the development of the European Parliament, the 1999 election should have shattered any such routine because, shortly before the election, the Parliament had done the unthinkable and forced the resignation of the Commission. The publicity attendant on this event was all embracing. This saga continued for weeks. The European Parliament seemed to be behaving at last like a real parliament by posing a significant threat to something akin to an executive. Yet, judged by the ultimate measure of the practical involvement of the voters, the 1999 elections that followed did not even rise to being routine. Instead of surging ahead on the back of the stimulus of the confrontation between the Commission and the Parliament and instead of being spurred on by the reformist zeal of the Parliament, voter turnout actually fell—from 58.5 to 52 percent across all the member states, and from 49.4 to 39.4 percent in those member states that did not have systems of compulsory voting. The turnout in 1999 confirmed and amplified a downward trend that extended over all five elections to the European Parliament. As the actual power of the Parliament has increased, turnout in elections to it has systematically and progressively decreased. Given the controversial political circumstances that preceded the election, the decline in turnout raises in a particularly insistent way the question of what was the mind or the mentality of the electorate in 1999 and how did it compare to 1994. In approaching this question, this chapter focuses on attitudes to the reliability of the Commission and of the Parliament, on perceptions of the power of the Parliament, and on perceptions of what is at stake in European Parliament elections.

Whatever may have happened to people's confidence in the Commission in the immediate aftermath of the so-called scandals and as a result of the mass resignation of the Commission in March 1999, the Commission did not suffer any long-term damage as far as public opinion was concerned. As figure 1.3 shows, the European public's sense of the reliability of the Commission was the same in the autumn of 1999 as it had been in midsummer 1994. Now it is possible that this was still a negative outcome: It may have been the case that prior to the 1999 crisis, the Commission had substantially improved its reliability rating with the public and that the autumn 1999 data in figure 2 represent a downturn. Since we lack precisely comparable data for the full intervening period, we cannot be certain of this one way or the other. What we can be certain of is that confidence in the ability of the Commission to make sure that "the decisions made by the European Union are in the interests of people like yourself" was no higher and no lower in late 1999 than it had been in mid-1994.

While this apparent resilience of the Commission may appear to be good news for European integration, the fact that the reliability rating of the Parliament was also virtually the same in autumn 1999 as it had been in mid-1994 is, from the point of view of supporters of the Parliament, disappointing. The Parliament had flexed its political muscle to an unprecedented degree. One would have expected that this would have produced some political returns in the form of an increase in the belief among the mass public that the Parliament was an effective guardian of the public's interests. Instead, what one finds is a remarkable degree of stasis. The Parliament's reliability rating in 1999 was for all practical purposes the same as it had been in 1994. Note that this is the case whether one compares the 1994 data to the 1999 data for the 12 older member states or for the EU 15 as a whole (see figure 1.3).

However, when attention is focused on perceptions of the *power* of the Parliament, things can be seen to have improved since 1994, at least to some extent. As figure 1.4 shows, the proportion of European citizens dismissing the Parliament as virtually powerless declined slightly—from 9 to 5 percent. In addition to this, the proportion giving the EP a "fairly low" power rating declined from 42 to 35 percent. Corresponding to these changes was a modest rise in the proportion that saw the Parliament as having a "fairly high" power rating (6 to 8 on a 10-point scale). The assessment of the gains evident in figure 1.4 must be tempered by three factors: First, the proportion of "don't knows" increased; second, the three member states that joined the Union after the 1994 elections made a disproportionate contribution to the

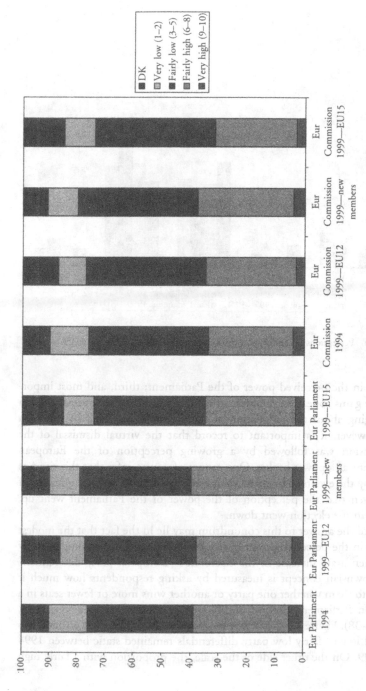

Figure 1.3 Perceptions of the Reliability of the European Commission and the European Parliament in Ensuring the Responsiveness of European Decision Making, 1994 and 1999

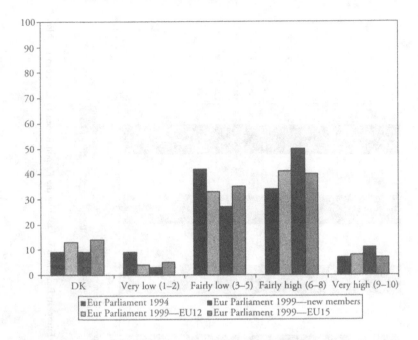

Figure 1.4 Perceptions of the Power of the European Parliament, 1994 and 1999

increase in the perceived power of the Parliament; third, and most important, the gains appear very modest when set against the extraordinary action
of bringing about the resignation of the Commission. That having been
said, however, it is important to record that the virtual dismissal of the
Commission was followed by a growing perception of the European
Parliament as a power broker. One cannot of course infer that this rise was
caused by the events in question. Nor can one easily answer the question of
why, given that the perception of the power of the Parliament went up,
turnout in the election went down.

Part of the answer to this conundrum may lie in the fact that the modest
increase in the perceived power of the Parliament was not accompanied by
a commensurate increase in the electorate's "European party differential."
This Downsian concept is measured by asking respondents how much it
matters to them whether one party or another wins more or fewer seats in a
European Parliament election (see Blondel, Sinnott, and Svensson, 1998,
pp. 137–38). Figure 1.5 shows that the proportions of European citizens who
reported low or fairly low party differentials remained static between 1994
and 1999. On the other side of the scale, the proportion with a fairly high

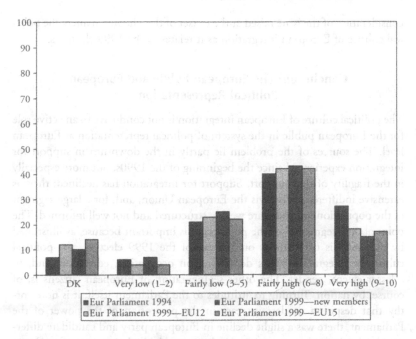

Figure 1.5 Party Differentials at European Parliament Elections, 1994 and 1999

differential also remained unchanged. In fact, the only statistically significant change over the five-year period is a small decline (of ten percentage points) in the proportion having a very high European party differential. In short, the modest gains made by the European Parliament with respect to its power profile did not translate into equivalent increases in the perception of what was at stake in the election, at least not in party terms.[2]

Analysis of the 1994 data shows that, even after considering a wide variety of other influences, European party and candidate differentials had a significant effect on abstention in the 1994 elections. The research also shows that perceptions of the power of the Parliament did not have any discernible impact on turnout (see Blondel, Sinnott, and Svensson, 1998, pp. 222–36). This is consistent with the findings presented above, that is, that the perceived power of the Parliament increased; that party and candidate differentials, to the extent that they moved at all, went down slightly; and that turnout also went down. These may be mere coincidences, however, and, before making any strong causal inferences based on these findings, a lot more analysis of the data in Eurobarometer 52 will be required.[3] In the meantime, armed with the evidence that we have presented and analyzed, we can return to the consideration of the issue raised at the outset of this chapter, namely the political culture of European integration as it relates to the 1999 elections.

Conclusion: The European Public and European Political Representation

The political culture of European integration is not conducive to an active role for the European public in the system of political representation at European level. The sources of the problem lie partly in the downturn in support for integration experienced since the beginning of the 1990s, but more especially in the fragility of that support. Support for integration has declined; there is extensive indifference vis-à-vis the European Union; and, for a large segment of the population, attitudes are weakly structured and not well informed. The point about "segments" of the population is important because, as illustrated by our analysis of attitudes on the eve of the 1999 election, the political culture of integration varies depending on occupation, education, and, to a lesser extent, gender. The other segmentation of European citizens is, of course, by nation. Turning to attitudes to the Parliament itself, it is noteworthy that despite a modest increase in the perception of the power of the Parliament, there was a slight decline in European party and candidate differentials, that is, in the perception that it mattered whether one set of parties or candidates won more or fewer seats in a European Parliament election. All of this is consistent with the substantial decline in turnout that occurred in 1999. More research will be required before one can be certain about the precise causal connections between these variables. However, at this stage one can be certain that the widespread abstention at the European Parliamentary elections and the political culture within which that abstention took place pose significant challenges to the pursuit of a closer involvement of citizens in the process of political representation at the level of the European Union.

Notes

1. The "very" was added to "relieved" in 1993. Unfortunately, this question is not asked as regularly as the others and has been particularly scarce in recent years.
2. While space does permit detailed treatment, it is worth noting that the same failure is evident if one looks at candidate differentials.
3. This analysis is already under way in the context of a large-scale study entitled "Democratic participation and political communication in systems of multi-level governance," which is sponsored by the EU Fifth Framework Programme for Research and Technological Development. The authors of the present chapter are participants in that project. The questions from EB 52 analyzed in this chapter were inserted on behalf of the project. Grateful acknowledgement is made to the

Press and Information Directorate of the European Parliament and to the Eurobarometer Unit in the Commission's Directorate General for Education and Culture for their support for this aspect of the project.

References

Blondel, Jean, Richard Sinnott, and Palle Svensson (1998), *People and Parliament in the European Union; Participation, Democracy and Legitimac* Oxford: Oxford University Press.

Bréchon, Pierre, Bruno Cautrès, and Bernard Denni, "L'évolution des attitudes vis-à-vis de l'Europe," in Pascal Perrineau, and Colette Ysmal, *Le vote des Douze. Les élections européennes de juin 1994.* Paris: Presses de Sciences Po, 1995, p. 175.

Cautrès, Bruno, "Les attitudes vis-à-vis de l'Europe," in Pierre Bréchon and Bruno Cautrès (eds.), *Les enquêtes Eurobaromètres. Analyse comparée des données socio-politiques.* Paris: L'Harmattan, 1998, pp. 91–113.

Cautrès, Bruno and Bernard Denni, "Les attitudes des Français à l'égard de l'Union européenne: les logiques du refus," in Pierre Bréchon, Annie Laurent, and Pascal Perrineau (eds.), *Les cultures politiques des Français.* Paris: Presses de Sciences Po, 2000.

Converse, Philip E. (1964), "The Nature of Belief Systems among Mass Publics" in *Ideology and Discontent*, ed. D. E. Apter. New York: Free Press.

Eichenberg, Richard C. (1998), "Measurement Matters: Cumulation in the Study of Public Opinion and European Integration," Paper presented at the American Political Science Association annual meeting, September 3–6, 1998, Boston.

Eurobarometer (1995), *Report on Standard Eurobarometer 43*, European Commission Directorate General X, autumn.

Franklin, Mark, and Cees van der Eijk (eds.) (1996), *Choosing Europe? The European Electorate and National Politics in the Face of Union.* Ann Arbor: University of Michigan Press.

Lindberg, L. N., and S. A. Scheingold (eds.) (1970), *Europe's Would-Be Polity: Patterns of Change in the European Community.* Englewood Cliffs, NJ: Prentice Hall.

Niedermayer, Oskar (1995), "Trends and Contrasts," in Oskar Niedermayer and Richard Sinnott (eds.), *Public Opinion and Internationalized Governance.* Oxford: Oxford University Press.

Reif, Karl Heinz, and Hermann Schmitt (1980), "Nine Second-Order Elections: A Conceptual Framework for the Analysis of European Elections Results," in *European Journal of Political Research*, 8: 3–4.

Sinnott, Richard (1995), "Policy Orientations and Legitimacy," in Oskar Niedermayer and Richard Sinnott (eds.), *Public Opinion and Internationalized Governance.* Oxford: Oxford University Press.

Sinnott, Richard (2000), "Knowledge and the Position of Attitudes to a European Foreign Policy on the Real-to-Random Continuum," in *International Journal of Public Opinion Research*, vol. 12, n. 2: 113–137.

CHAPTER 2

Dynamic Representation: The Case of European Integration*

Hermann Schmitt and Jacques Thomassen

Introduction

This chapter raises two questions: Why are party voters less favorable toward specific EU policies than are party elites? Second, how does political representation of EU preferences actually work? Is it an elite- or a mass-driven process? The data sets of the European Election Studies 1979 and 1994 are analyzed, which involve both an elite and a mass survey component. In contrast to earlier research, it appears that political representation of EU preferences works rather well regarding the grand directions of policymaking, and that party elites behave responsively in view of changing EU preferences among their voters.

Political Representation in the European Union: What We Know About It

Empirical investigations into the effectiveness of political representation in the European Union are scarce.[1] The relevant literature includes part of the work of the European Election Studies research group (mostly van der Eijk and Franklin, 1991 and 1996; and the contributions to Marsh and Norris, 1997), a few other studies based on Eurobarometer (Niedermayer and Sinnott, 1995; Blondel et al., 1998), and data on party manifestos (Carrubba, n.d.). Recent additions to these empirical investigations of EU democracy are the results of the European Representation Study 1994–1997, published in two companion volumes (Schmitt and Thomassen, 1999; Katz and Wessels, 1999).

It is a complex undertaking to assess the effectiveness of political representation in the multitiered polity of the European Union. Depending on the policy area concerned, EU governance oscillates between an intergovernmental and a supranational mode. Due to this, the European Representation Study was designed to investigate the preconditions and effectiveness of electoral representation regarding both European elections and first-order national elections. (Nonelectoral mechanisms of political representation such as lobbying, while arguably of particular importance at the EU level, could not be considered.) The criteria being tested for these two channels of electoral representation are derived from the "responsible party model." This model assumes that competitive and cohesive parties exist; that voters have policy preferences and perceive the policy options on offer correctly; and that voters in the end base their electoral choice on these preferences. If these conditions are met, the process of political representation should result in a close match between the preferences of party voters and the policies of party elites.

Large-scale representative surveys among the mass publics and among members of the European Parliament and of national parliaments were conducted to assess the validity of these assumptions. The results can be summarized as follows: EU party elites are no less cohesive than national party elites. Voters hold policy preferences. They also recognize where the parties stand with regard to the grand lines of policymaking, while the more detailed EU policy positions of political parties escape many voters. It is hardly surprising, then, that those EU policies are largely irrelevant for the vote, while general policy views (as expressed, e.g., in terms of left and right) are significantly related to it.[2]

As a consequence, political representation in the European Union works fairly well as far as general policy views are concerned; if it comes to the specifics of European Union policymaking, the congruence between voters and their representatives is remarkably poor. Political elites are much more European-minded than their voters regarding questions such as the abolition of border controls or the elimination of national currencies in favor of a new common European currency. It is striking that this representative deficiency is not specific to the EU channel of electoral representation. National representatives are no less European-minded than their colleagues in the European Parliament, and are thus equally distant from their voters on these specific EU policies.

Compared to the results of earlier work, these findings evoke a number of further questions. One is whether the apparent ineffectiveness of political representation with regard to EU policies is caused by the "Europeanness" of these issues or by their specificity. Phrased in somewhat less obscure terms, the question is whether voters are less integrationist than their representatives, or whether they are simply less expert—less informed and hence more afraid of changes in the status quo. Relying on indicators of general EU approval, van der Eijk and Franklin (1991) found a rather close match between voters' orientations and their perceptions of where the parties stood. While this seems to suggest that voters are no less European-minded than their representatives, it could well be a result of wishful thinking of party voters (van der Brug and van der Eijk, 1999) rather than an adequate account of reality. A more definitive answer to this question obviously needs to compare original measures taken from party voters and party elites.

Another question goes beyond the *whether* and asks *how* political representation works in the European Union. There are two competing views about this, one elite- and the other mass-driven. According to the elite-driven view, attitudes and preferences of voters tend to follow the lead of political elites and political events more generally (e.g., Page and Shapiro, 1992). In the mass-driven perspective, political elites behave responsively vis-à-vis changes they perceive in the attitudes and preferences of their voters (e.g., Stimson, 1991).

The standard view on European integration is that it is largely an elite-driven process (Deutsch, 1968; Wessels, 1995) that rests upon the permissive consensus of the general public (Lindberg and Scheingold, 1970). This coincides with findings of national representation studies, which cover a broader range of issues (Esaiasson and Holmberg, 1996; Holmberg, 1997). More recent irritations among the mass publics following the Maastricht process have cast doubts on this view (Niedermayer, 1995). And a forthcoming diachronic analysis of voter attitudes about EU membership and party elite positions toward the EU suggests that party elites are responsive to changing voter orientations rather than the other way around (Carrubba, n.d.). However, a conclusive answer to this question needs to be based on a dynamic model built on comparable indicators for both mass and elite levels and for at least two points in time. This is what we try to do in this paper. The results will have important implications not only for our views on the process of European integration; they will be relevant also for our understanding of how democracy works in more general terms.

Research Strategy, Data, and Indicators

Answering our first research question involves comparing measures of mass–elite congruence for issues of different specificity. Units of analysis are not individual candidates and the electorate of their constituency, as the individualistic model of the American Representation Study would have it (Miller and Stokes, 1963), but national aggregations of party elites and party voters—so-called party dyads according to the responsible party model

(Holmberg, 1974; Thomassen, 1976; Dalton, 1985; Schmitt, 2001). The older individualistic conception of political representation does not capture the reality of modern European party democracies, which is shaped by political parties much more than by individual candidates. The standard—but not undisputed[3]—measure of congruence between aggregate positions of party voters and party elites is the correlation coefficient.[4]

An answer to our second research question requires a "dynamic" analysis. Any dynamic analysis of processes of political representation necessitates at least two observations in time measured at two levels. "Observations" refers to positions on one or more relevant dimensions of political competition and controversy. For the European Union, these are the left–right dimension, which structures party competition in its constituent national polities, and the integration-independence dimension, which might be more salient for the EU polity (e.g., Hix, 1999; Marks and Wilson, 1999). "Level" refers to the two groups of actors involved—the represented (i.e., voters) and the representatives (i.e., party elites).

Given a research design that allows us to control variation in voters' and elites' positions for two points in time, one could, on each level, utilize the earlier observation as a predictor of the present and model the relationship between voters' and elites' present positions in a nonrecursive way. This can be graphically displayed as follows (figure 2.1).

We will test this model with data from the European Election Studies 1979 and 1994. Large-scale election studies were fielded in both election years.[5] A comparable core is formed by two basic questions put in identical

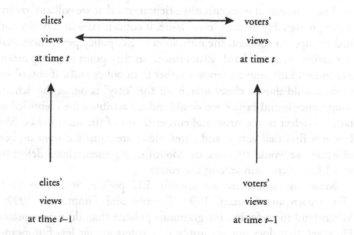

Figure 2.1 The Basic Research Design

wording in both studies to party voters and party candidates.[6] One is the well-known left–right self-placement question; the other is a basic measure of integration (vs. independence) preferences.[7] This latter instrument is also a perfect tool with which to investigate our first research question, to which we now turn.

Why Are Voters More Skeptical about EU Policies than Are Party Elites?

Voters are consistently, and considerably, more skeptical about integrationist policies than their representatives are (Schmitt and Thomassen, 1999). Is this deficiency in the system of political representation of the European Union a result of a structural conservatism on the part of party voters, or of a deliberately less integrationist stance? We have already pointed out that voters might be more conservative than political elites because they are not as well informed about the likely consequences of a particular policy, thus face higher decisional insecurity and are more likely to prefer what they have rather than to opt for a change—in whatever direction this change would actually lead. Such structural conservatism should be apparent in view of any major policy change and should not be specific to the policies of European integration. It should not, however manifest itself if voters are confronted with "easy," rather than "hard," issues (Carmines and Stimson, 1980), which focus on policy ends rather than policy means. For those issues, the relative lack of information should not lead to decisional insecurity, and structural conservatism should not be apparent.[8]

The indicator of integrationist orientations that we will analyze in the following is arguably such an "easy" issue. It concentrates on a policy end: a unified Europe. In contrast, the indicators of EU policy preferences analyzed in our earlier work are "hard" issues insofar as they point to policy means—e.g., a common European currency—rather than policy ends. If voters' and elites' views should show a closer match for the "easy" issue, as the deficient issue congruence found earlier, we would indeed attribute the identified representational defect to the structural conservatism of the mass public. Should we however find that voters' and elites' views are equally distant on both types of issues, we would attribute the identified representational defect to an elevated Euroskepticism among the voters.

Mass–elite agreement on specific EU policies is known to be poor (Thomassen and Schmitt, 1997; Schmitt and Thomassen, 1999: ch. 9). Voters tend to prefer less integrationist policies than do their representatives. However, that does not yet imply that voters are far less European-minded than their representatives. They might be merely insecure about the outcomes of particular EU policies and therefore tend to prefer what they perceive the status quo to be. This is exactly what we find. Our analyses reveal that while their policy preferences diverge, integrationist orientations of voters and party elites match about as closely as their left–right orientations, which is a very close match indeed (table 2.1). Political representation in the European Union might be deficient as regards the specifics of EU policymaking, but it seems to function well as far as the grand directions of public policy are concerned. Integrationist orientations of voters are well represented by their party elites.

This is not to say that voters are as integrationist, or pro-European, as their representatives. Figures 2.2 and 2.3 show that party voters, while not far from the views of the elite of the party they voted for, are systematically somewhat less integrationist. This is a consolidated finding, as it holds for both the 1979 and 1994 studies. Only a small minority of parties figure below the diagonal. These are those whose voters are *more* integrationist than the party elite. For the majority of parties, the contrary is true. Party elites are somewhat more "European" than the mass public, representatives

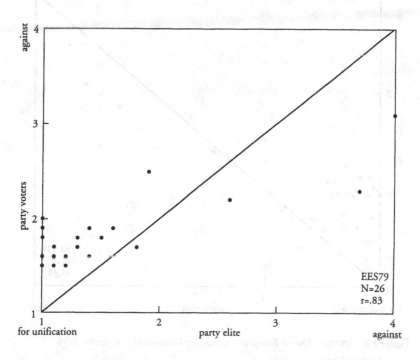

Figure 2.2 Attitudes Toward European Unification: Voter–elite Congruence, 1979
Source: European Election Study 1979.

Table 2.1 Voters are insecure regarding EU policies rather than opposed to European integration

		Open borders	Employment program	Common currency	European unification	Left–Right
1979	Pearson's r	n.a.	n.a.	n.a.	.83	.88
	valid cases				26	26
1994	Pearson's r	.56	.52	.47	.84	.82
	valid cases	46	46	46	46	46

Source: Voters and candidates surveys of the European Election Studies 1979 and 1994. n.a. = not ascertained. Correlated are elite and voter positions. Units of analysis (= cases) are party dyads. A party dyad exists if a reliable positional measure for both voters and party elites (i.e., party candidates) is available. Positional measures are arithmetic means. These measures are considered reliable if they are based on the voters side on at least 20 voter interviews, and on the elite side on at least 5 candidate interviews.

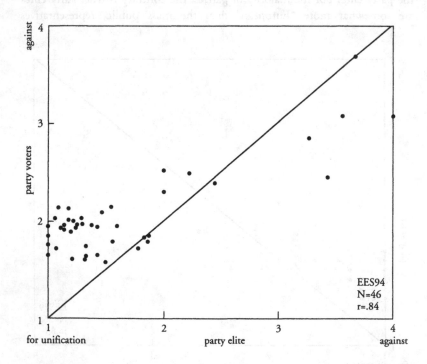

Figure 2.3 Attitudes Toward European Unification: Voter–elite Congruence, 1994
Source: European Election Study 1994.

somewhat more integrationist than their voters. This is a relevant piece of information, even if the discrepancies are modest.

A similar phenomenon is known to exist for left–right orientations. Correlation coefficients indicate that voters and party elites place themselves very close to one another on this dimension. However, a more detailed inspection reveals a small but systematic difference: Elites are regularly somewhat more to the left than their voters. The opposite is observed only for a few parties on the far right (figures 2.4 and 2.5). There is hardly any empirical representation study that does not report on this phenomenon (e.g., Converse and Pierce, 1986). The standard explanation refers to the differences in the social status between representatives and represented, and to the different values that originate in different educational and professional careers and in different social environments more generally.

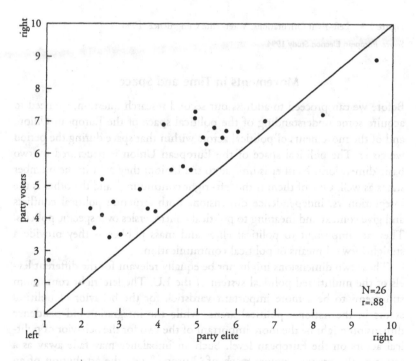

Figure 2.4 Left–right Orientations: Voter–elite Congruence, 1979
Source: European Election Study 1979.

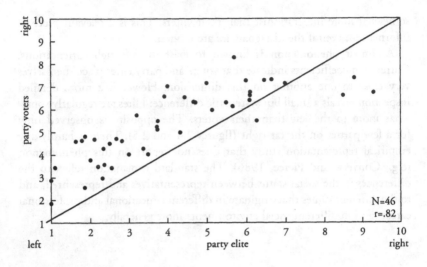

Figure 2.5 Left–right Orientations: Voter–elite Congruence, 1994
Source: European Election Study 1994.

Movements in Time and Space

Before we can proceed to address our second research question, we need to acquire some understanding of the political space of the European Union, and of the movements of political actors within that space during the period we cover. The political space of the European Union is structured by two basic dimensions. Neither is unique to the Union; they exist in the member states as well. One of them is the left–right continuum,[9] and the other is the integration vs. independence dimension. Both structure political conflicts and give context and meaning to political controversies over specific policies. They are important to political elites and mass publics as they provide a straightforward means of political communication.

These two dimensions might not be equally relevant for the different levels of the multitiered political system of the EU. The left–right continuum still seems to be a more important yardstick for the behavior of political actors in the national political arena, while the integration-independence dimension might be the more important of the two for the behavior of political actors on the European level. Such an imbalance may fade away as a result of the growing policy reach of "Europe," i.e., the attribution of an increasing number of common-concern issues (as opposed to EU constitutional issues) to the government at the European level of the multitiered political system of the EU. But this is not a central argument for the following and it may suffice to say that the weight of these two dimensions is not fixed but variable across levels and over time.

There is one additional complication that we need to address. The left–right and integration-independence dimensions might be connected. The more they are correlated, however, the less it is justified to consider them both. If, for example, integration-independence preferences could be predicted reasonably well from left–right positions, it would be irrelevant to investigate the former. A more mundane aspect of this is that the conventional orthogonal arrangement of the two dimensions (e.g., Hix and Lord, 1997; Hix, 1999; Marks and Wilson, 1999) would be inadequate.

Comparing findings from the European Election Studies of 1979 and 1994, it seems that the integration-independence dimension was liberated from the left–right frame during the 1980s and 1990s. Still in 1979, integrationist views were "right wing." While this was least visible among the mass public, it was quite pronounced among political elites (table 2.2). Aggregating elite responses at the level of national parties, and particularly at the level of European parties/Parliamentary party groups, amplified this phenomenon. In the first directly elected European Parliament of 1979, integrationists sat on the right of the assembly, and protagonists of national independence sat on the left.

The Euroskepticism of the left has been melting away. In 1994, left-wingers were even somewhat more pro-European than right-wingers. This is evident for both voters and political elites. However, while in 1979 national parties and, in particular, EU parties and EP groups amplified individual tendencies through the aggregation of like-minded men, national parties and EP groups seem to have lost this capacity since then. In 1994, left-wingers were somewhat more integrationist everywhere—within the SPD and CDU,

Table 2.2 Euroskepticism of the left is melting away (Correlations between European integration and left–right orientations at different levels and times)

		EP groups[a]	National parties[b]	Individual EP candidates	Individual voters[c]
1979	Pearsons r	−.74[d]	−.48	−.36	−.04
	valid cases	10	26	611	6881
1994	Pearsons r	+.27	+.12	+.15	+.05
	valid cases	9	54	744	10385

Source: European Election Studies 1979 and 1994. [a]EP candidates are aggregated into EP groups that they (would) have joined after election, and group means are correlated; [b]EP candidates are aggregated according to national party membership/affiliation, and the party means are correlated (parties with less than five interviewed candidates are disregarded); [c]individual respondents are weighted to improve representatively according to demographic characteristics within nations and to relative national population size at the EU level; [d]read: the more to the left, the less "for" unification.

PvdA and CDA, PES and EPP. This is compatible with Marks and Wilson's vision of a "Social Democracy valley" (Marks and Wilson, 1999: 117 ff). According to them, Social Democrats (and left-wingers more generally) first opposed European integration as it meant a loss of national control and ability to steer the economy and the welfare state. After a while, however, the ongoing processes of market globalization made them realize, according to these authors, that the only chance to regain control over market forces was to accept the process of supranationalization and compete for control over EU government. In a nutshell, Social Democrats/left-wingers made their peace with Europe when they realized that the nation-state was no longer a suitable framework for (post-)Keynesian policies.

But European integration in 1994 was not as left wing as it had been right wing in 1979. The two basic dimensions of the European political space had gained greater independence from one another over these years. We can thus determine the positioning and movements of relevant political actors in the political space that is defined by these two orthogonally arranged dimensions. Relevant actors are political parties. National parties form national governments, which together constitute the European Council; and they align within European parties and the political groups of the European Parliament.

There are different ways to determine party positions in the EU political space. Earlier studies used voter positions (Hix and Lord, 1997), derived elite positions from content analyses of party documents (Hix, 1999), or based their measurement on expert judgments (Marks and Wilson, 1999). Published results do not coincide and are therefore not very reliable. We use another measure. As party positions are defined by party elites, we will exploit our two candidate surveys as the most immediate source for estimating party positions and their changes over time. In this descriptive analysis, we concentrate on European parties as represented in the group structure of the European Parliament.[10] National parties will be our focus in the causal analysis that follows. Figure 2.6 displays the positions of European Union parties and parliamentary groups and how they have changed as represented in the group structure of the European Parliament. Two developments stand out, and both occurred at the poles of the left–right continuum rather than in the political center. One is the growth of anti-integrationism on the extreme right. Still in 1979, this group was among the most pro-European; 15 years later, they were the second most anti-European group in the European Parliament.

The other major change took place on the left. This one is a bit more complex because of contemporary history: The breakdown of Communism led to a major restructuring of the European party system after 1989. Former Communist Parties changed camps and are now aligned with the Socialist group (the Italian Democratici di Sinistra being an example). Orthodox

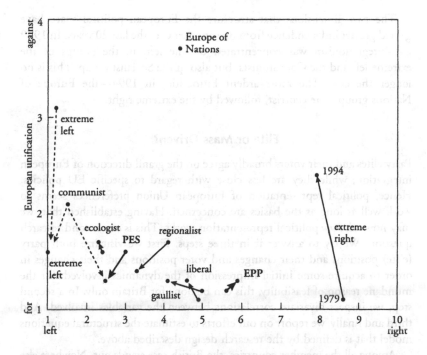

Figure 2.6 Movements in Time and Space: EP Groups, 1979–1994

Source: Candidate surveys of the European Election Studies 1979 and 1994. EP candidates are aggregated into EP groups that they (would) have joined after election, and group means are displayed. The arrows for the two predominant groups—the Party of European Socialists (PES) and the European People's Party (EPP)—are bold, while the dotted arrows signify movements of group components (rather than those of entire groups).

Communist splinters (Italy providing again a good example with the Rifondazione Comunista) and small Communist standpatters (like the Parti Communiste Français) today constitute what once was called the Communist group. They incorporate some of the former extreme left that disappeared as a group (or rather, merged with orthodox Communists). As a result of these developments, both the Socialist and the Communist/extreme-left group moved somewhat to the left. But this is not the most spectacular development to mention here. Much more pronounced are the movements of both groups in the pro-European direction. Regionalists also became more integrationist, while ecologists, Liberals, Gaullists, and the European People's Party became somewhat more skeptical about European integration. With regard to the EPP, this might reflect a composition effect more than anything else: In our 1994 survey, this group incorporated conservative parties (the British Conservatives in particular), which in 1979 were still on their own (not shown).

The two dimensions that structure the European political space have gained greater independence from one another over the last 20 years. In 1979 anti-integrationism was concentrated on the left, in the groups of the extreme left and the Communists, but also in the Socialist group. This is no longer the case. The most ardent Eurocritics in 1994—the Europe of Nations group—are centrist, followed by the extreme right.

Elite or Mass Driven?

Party elites and their voters broadly agree on the grand direction of European integration, while they are less close with regard to specific EU policies. Hence, political representation of European Union preferences seems to work well as long as the basics are concerned. Having established that, we may now ask *how* political representation works. This is our second research question. We try to answer it in three steps. First we visualize both party (elite) positions and their changes and voter positions and their changes in order to acquire some initial impression of the dynamics involved; for the mundane reason of feasibility, this can be done for Britain only. In a second step, we inspect bivariate correlations between the variables involved. And third and finally, we report on our efforts to estimate the structural equations model that is defined by the research design described above.

Among all the member countries, the British case stands out. Nowhere else are party movements as pronounced as in Britain. Concentrating on voters' orientations, students of British electoral politics have taken note of this (e.g., Evans, 1998; 1999). However, party elites are moving faster, or rather farther, than their voters (figure 2.7). If we focus on them for a moment, we realize that movements on the integration-independence dimension are more extensive than those on the left–right dimension. In 1979, the Conservatives took a clear integrationist position, while Labour elites were still very skeptical about European integration. The reverse is found for 1994, when Labour took the lead toward further unification, while Conservatives opposed it. It is astounding how far these shifts in party elite positions go. Nowhere else do we find a similar phenomenon. There was also some elite movement on the left–right dimension in that period. The Liberals went to the left, as did Labour, while Conservatives went to the right. We thus observe among British party elites a left–right polarization, in addition to the change of roles on the European question.

It seems impossible on the basis of the visual inspection of these three cases to decide whether voters follow elites or elites are responsive to voters. We need to use other techniques of data analysis and consider additional information in order to come closer to answering our research question. Table 2.3 displays, for voters' and elites' 1994 positions on both the

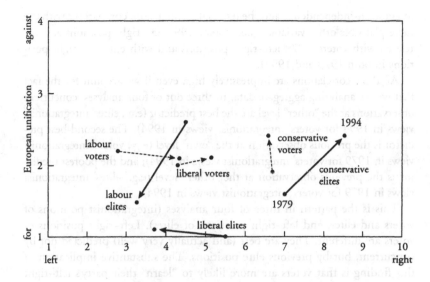

Figure 2.7 Changes of Political Orientations of Voters and Elites of Major British Parties, 1979–1994

Source: Voters and candidates surveys of the European Election Studies 1979 and 1994. Voters are much less mobile, and probably more important to our research question here, they do not necessarily follow the directions their party elites seem to give. While Labour elites become more integrationist and left wing, Labour voters move to the right and stay where they are regarding Europe. Conservative voters follow their elites—or lead them—in a more Eurocritical direction, while staying in the center rather than becoming more right wing. Liberal voters, finally, are farthest from their most integrationist elite and move to the right while their party goes left.

Table 2.3 Determinants of voters' and elites' views in 1994 (correlation and determination coefficients)

		Elites 1994			Voters 1994		
			r	*r²*		*r*	*r²*
European	Best predictor	Voters 94	.895	.801	Elites 94	.895	.801
Integration	2nd best	Elites 79	.859	.738	Voters 79	.887	.787
	Poorest	Voters 79	.842	.709	Elites 79	.821	.674
Left–right	Best predictor	Voters 94	.807	.651	Elites 79	.954	.910
	2nd best	Elites 94	.788	.621	Voters 79	.851	.724
	Poorest	Voters 79	.735	.540	Elites 94	.807	.651

Source: European Election Studies 1979 and 1994. N of cases (parties) = 18. All coefficients significant better than .01 (two-tailed tests). The parties for which elite and mass data are available for both 1979 and 1994 are: CVP (B), SP (B), PVV (B), SD (DK), V (DK), FbmEF (DK), CDU (D), SPD (D), PS (F), UDF (F), MSI/AN (I), DC/PPI (I), CDA (NL), PVDA (NL), VVD (NL), Conservative Party (GB), Labour Party (GB), and Liberal Party/Liberal Democrats (GB).

integration-independence and the left–right dimension, correlations with the three plausible other variables (i.e., voters' 1994 left–right positions are correlated with voters' 1979 left–right positions and with elite left–right positions in both 1979 and 1994).

All these correlations are impressively high even if we account for the fact that we are analyzing aggregate data. In three out of four analyses, concurrent observations at the "other" level are the best predictor (e.g., elites' integrationist views in 1994 for voters' integrationist views in 1994). The second-best predictor is the previous observation at the "own" level (e.g., voters' integrationist views in 1979 for voters' integrationist views in 1994); and the poorest predictor is the previous observation at the "other" level (e.g., elites' integrationist views in 1979 for voters' integrationist views in 1994).

This is the pattern in three of four analyses (integrationist positions of voters and elites, and left–right positions of elites). Left–right positions of voters are different. They are best (and actually very well) predicted not by concurrent, but by previous elite positions. The substantive implication of this finding is that voters are more likely to "learn" their party's left–right position at some earlier point—in a politically formative phase—than to adapt to the party's current position (or the other way round: that the party would adapt to the voters' current position). We will come back to this issue and its meaning for models of political representation and voting behavior in greater detail elsewhere. For present purposes it may suffice to say that the mutual determination of voters' and elites' current orientations is stronger for the integration-independence dimension than for the left–right dimension, and that the basic causal structure of our research design is more appropriate for integrationist views than for left–right positions.

We therefore concentrate our final analysis on the integration-independence dimension. Figure 2.8 displays standardized regression estimates for the nonrecursive relationship between integrationist views of party elites and of party electorates. The model is not fitted to the data, as we are not interested in accounting for every trace of covariance among the variables. What is essential here are stable and reliable estimates for the two central effects, from current elite views to current voter views and vice versa. These indicate that the voter impact on elite views (bottom-up effect) is more substantial than the elite impact on voters' views (top-down effect).

This nicely supports the results of Carrubba's analysis of what he calls "the electoral connection in EU politics." Relating average EU membership evaluations of potential party electorates around election time on the one hand to EU mentions in party manifestos on the other hand, he reports a significant effect of voter positions on party elite behavior (the content of election manifestos). This is exactly what we find—voters' views have a somewhat greater impact on

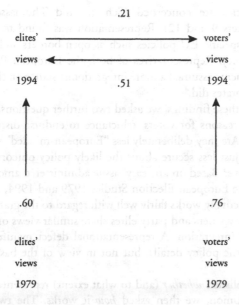

Figure 2.8 Representation from Below: A Simple Causal Model Linking Elites' and Voters' Views on European Integration in 1994 and 1979

Source: European Election Studies 1979 and 1994. N of cases (parties) = 18. The parties for which elite and mass data are available for both 1979 and 1994 are: CVP (B), SP (B), PVV (B), SD (DK), V (DK), FbmEF (DK), CDU (D), SPD (D), PS (F), UDF (F), MSI/AN (I), DC/PPI (I), CDA (NL), PVDA (NL), VVD (NL), Conservative Party (GB), Labour Party (GB), and Liberal Party/Liberal Democrats (GB). The model has been estimated with EQS.

elite orientations than vice versa. However, it seems to be at odds with what we know about political representation in national polities. Esaiasson and Holmberg (1996) and Holmberg (1997) find for a majority of issues that Swedish voters adapt their views to elite positions with some time lag. It may well be that the direction of determination depends on the nature of the issue involved. We know little about these subtleties below the surface of issue congruence. One of the reasons is that suitable data sets for testing somewhat more elaborate hypotheses about cause and effect in representational relationships are very rare.

Summary

This chapter took one of the main results of the European Representation Studies as its starting point. According to this, the system of political representation in the European Union seems to work well as long as other than

European policies are concerned (Schmitt and Thomassen, 1999, in particular chapters 9 and 12). Representation was found to be ineffective with regard to specific EU policies such as open borders and the common European currency. Representatives of both the European Parliament and national parliaments assumed a more integrationist stance on those questions than their electorates did.

Building on these findings, we asked two further questions. The first had to do with the reasons for voters' reluctance to endorse distinctly integrationist policies: Are they deliberately less "European-minded" than their representatives, or just less secure about the likely policy outcomes and hence more conservative? Based on an "easy" issue administered among voters and elites in both the European Election Studies 1979 and 1994, we found that representation actually works fairly well with regard to the grand direction of EU politics: Party voters and party elites share similar views on the question of more or less integration. A representational defect manifests itself only with regard to the policy details, but not in view of the basic question of where to go.

Having established *whether* (and to what extent) representation works in the European Union, we then asked *how* it works. The two dimensions structuring the European political space—left–right and integration–independence—have become more independent from 1979 to 1994; the Euroskepticism of the left has been melting away, and the most ardent anti-integrationists today are located in the center (and on the right) of the political spectrum. Analyzing integrationist and left–right orientations measured with identical instruments at mass and elite level at different points in time, we were able to show that current integrationist views of voters and elites greatly determine one another, while left–right orientations of voters are most strongly shaped not by current but by former elite positions. This suggests an explanation of acquisition and change in mass left–right orientations that focuses on socialization rather representation processes. We will explore this issue further in future work.

With regard to integrationist views, we found that voters seem to have a somewhat stronger impact on party elites than vice versa. Representation from above (Esaiasson and Holmberg, 1996) is obviously not the only mechanism at work in the complicated process of sociopolitical linkage and democratic decision making. While it seems to concur with what we find for left–right orientations, it does not fit our results with regard to integration vs. independence preferences. For the time being, we do not understand very clearly why this is so. Further research will be needed to shed more light in these dark corners of sociopolitical linkage.

Notes

* A version of Chapter 2 was previously published in *European Union Politics* (2000), vol. 1, no. 3: 299–318.

1. Data collections of the European Election Study 1979 were supported by grants from the Volkswagen Foundation, the European Parliament and the European Commission, and various research institutes; data collections of the European Election Study 1994 were supported by the German Research Foundation DFG, the Dutch Science Foundation NWO, and various research institutes. All these grants are gratefully acknowledged. In the course of the present research, Evi Scholz graciously supported us by making the 1979 candidates survey data accessible. This is also gratefully acknowledged. All data analyzed in this paper can be obtained for further analysis from the *Zentralarchiv für Empirische Sozialforschung* at the University of Cologne, Germany.

2. Issue effects on the vote are more pronounced for issue competence attributions than for parties' policy positions. This is consequential also for the measurement of political representation. Following the competence logic, measures of issue congruence should be based on issue salience rather than issue positions. A close match between voters' and elites' views is then indicated by similar salience evaluations rather than distances in their policy positions (Schmitt, 2001). The data sets analyzed in this article do not contain measures of salience attributions toward European integration, but measures of voters' and elites' positions on this question. These positional indicators, however, belong to the class of "easy" issues, which are closer to the vote and thus a better indicator of political representation than are "hard" issues (Carmines and Stimson, 1980).

3. Achen (1977, 1978) issued strong warnings against correlational measures of issue congruence that under certain distributional conditions could produce systematically distorted results. We agree with Converse and Pierce (1986, pp. 603 and 964–965) and many others that these conditions are hardly ever met.

4. In a strict sense, the simple fact of a general agreement between voters and party elites does not yet prove that the system of political representation is working. The congruence that we find in the issue positions of voters and party elites could be the result of systematic misperceptions. A full investigation of the effectiveness of the representational system would require determining the extent to which voters have a good understanding of where the party elites stand and vice versa. While we were able to study this in detail in our larger enquiry (Schmitt and Thomassen, 1999), data limitations prevent us from addressing those questions in the present diachronic analysis. However, we found in our larger study that issue congruence and perceptual accuracy go hand in hand. Thus we maintain that our results point in the right direction, however incomplete our present research design may be.

5. The EES 79 was essentially a study of political mobilization. It was made up of a voter survey, a campaign study, a middle-level party elite survey, and a candidates survey. The questions of the voter study could be added to the questionnaires of the Eurobarometer 11 (pre-electoral survey). The EES 94 was designed as a

representation study. It consisted of a voter survey, a candidates survey, a survey among MEPs, and another among members of national parliaments. The questions of the voter survey could be added to the questionnaires of Eurobarometer 41.1 (postelectoral survey).

6. Candidates are not representatives, at least not all of them. They are, however, part of the party elite more broadly defined. As we do not relay on an individualistic model of political representation (in which individual representatives do make a difference) but on a collectivist model (in which the party elite *in total* counts), the assumption that candidates' political views are a valid indicator of the partyelite more generally is probably not too heroic. See similarly Dalton (1985) and Marsh and Norris (1997).

7. The left–right question reads as follows: "In political matters people talk of the 'the Left' and 'the Right.' How would you place your views on this scale?" The answer scale ranges from 1 = left to 10 = right. The integration vs. nationalism question goes as follows: "In general, are you for or against efforts being made to unify western Europe?" Answering categories are: for–very much; for–to some extent; against–very much; against–to some extent. The integration question was asked early on in the Eurobarometer surveys as one of the four basic indicators of European attitudes.

8. The work of Converse (1964) and Zaller (1992) is probably most relevant when it comes to models of attitudes and attitude change. However, both are more interested in the stability and change of mass opinions than in systematic differences in political orientations of elites and the citizenry at large. While they therefore do not identify anything like the structural conservatism we are proposing, their general reasoning seems to support our claim.

9. The left–right dimension arguably is not one single ideological dimension, but rather a conglomerate of an economic (interest-based) and a cultural (value-based) subdimension (see e.g., Hix, 1999). While we ourselves have identified these subdimensions in an earlier work (Thomassen and Schmitt, 1999), we feel justified in the present analysis in concentrating on the one overarching left–right structure.

10. EP group affiliations of national parties are documented for 1979 in Reif and Schmitt (1980) and for 1994 in Hix and Lord (1997).

References

Achen, Christopher H. (1977), "Measuring Representation: Perils of the Correlation Coefficient," *American Political Science Review* 68: 805–15.

Achen, Christopher H. (1978), "Measuring Representation," *American Journal for Political Science* 22: 475–510.

Blondel, Jean, Richard Sinnott, and Palle Svensson (1998), *People and Parliament in the European Union*. Oxford: Clarendon Press.

Carmines, Edward G., and James A. Stimson (1980), "The Two Faces of Issue Voting," *American Political Science Review* 74: 78–91.

Carrubba, Clifford J. (n.d.), "The Electoral Connection in European Union Politics," *Journal of Politics* (forthcoming).

Converse, Philip E. (1964), "The Nature of Belief Systems in Mass Publics," in David Apter (ed.) *Ideology and Discontent*, pp. 206–261. New York: Free Press.

Converse, Philip E., and Roy Pierce (1986), *Political Representation in France*. Cambridge and London: The Belknap Press of Harvard University Press.

Dalton, Russel J. (1985), "Political Parties and Political Representation. Party Supporters and Party Elites in Nine Nations." *Comparative Political Studies* 18: 267–99.

Deutsch, Karl W. (1968), *Analysis of International Relations*. Englewood Cliffs, NJ: Prentice Hall.

Esaiasson, Peter, and Sören Holmberg (1996), *Representation from Above*. Aldershot, Hants: Gower.

Evans, Geoffrey (1998), "Euroscepticism and Conservative Electoral Support: How an Asset Became a Liability." *British Journal of Political Science* 28: 573–90.

Evans, Geoffrey (1999), "Europe: A New Electoral Cleavage?," in Geoffrey Evans and Pippa Norris (eds.), *Critical Elections*, pp. 207–222. London: Sage Publications.

Hix, Simon (1999), "Dimensions and Alignments in European Union Politics. Cognitive Constraints and Partisan Responses," *European Journal of Political Research* 35: 69–106.

Hix, Simon and Christopher Lord (1997), *Political Parties in the European Union*. London: Macmillan.

Holmberg, Sören (1997), "Dynamic Opinion Representation," *Scandinavian Political Studies* 20: 265–83.

Holmberg, Sören (1974), *Svenska Väljare*. Stockholm: Liber.

Katz, Richard, and Bernhard Wessels (1999), *The European Parliament, the National Parliaments, and European Integration*. Oxford: Oxford University Press.

Lindbergh, Leon N., and Stuart A. Scheingold (1970), *Europe's Would-Be Polity. Patterns of Change in the European Community*. Englewood Cliffs, NJ: Prentice Hall.

Marks, Gary, and Carole Wilson (1999), "National Parties and the Contestation of Europe," in Thomas Banchoff and Mitchell P. Smith (eds.) *Legitimacy and the European Union*, pp. 110–132. London and New York: Routledge.

Marsh, Michael, and Pippa Norris, eds. (1997), *Political Representation in the European Parliament*. Dordrecht: Kluwer [special issue of the *European Journal of Political Research*, vol. 32/2].

Miller, Warren, and Donald Stokes (1963), "Constituency Influence in Congress." *American Political Science Review* 57: 45–56.

Niedermayer, Oskar (1995), "Trends and Contrasts," in Oskar Niedermayer and Richard Sinnott (eds.) *Public Opinion and Internationalized Governance* [Beliefs in Government series, vol. 2], pp. 53–72. Oxford: Oxford University Press.

Niedermayer, Oskar, and Richard Sinnott, eds. (1995), *Public Opinion and Internationalized Governance*. [Beliefs in Government series, vol. 2]. Oxford: Oxford University Press.

Page, Benjamin I., and Robert Y. Shapiro (1992), *The Rational Public*. Chicago and London: University of Chicago Press.

Relf, Karlheinz, and Hermann Schmitt (1980), "Nine Second-Order Elections: A Conceptual Framework for the Analysis of European Elections Results," *European Journal of Political Research* 8: 3–44.

Schmitt, Hermann (1998), "Issue-Kompetenz oder Policy-Distanz?" in Max Kaase and Hans-Dieter Klingemann (eds.), *Wahlen und Wähler*, pp. 145–172. Opladen: Westdeutscher Verlag.

Schmitt, Hermann (2001), *Politische Repräsentation in Europa*. Frankfurt: Campus.

Schmitt, Hermann, and Jacques Thomassen, eds. (1999), *Political Representation and Legitimacy in the European Union*. Oxford: Oxford University Press.

Stimson, James A. (1991), *Public Opinion in America*. Boulder: Westview Press.

Thomassen, Jacques (1976), *Kiezers and Gekozenen in en Representative Democratie*. Alphen an den Rijn: Samsom.

Thomassen, Jacques, and Hermann Schmitt (1997), "Issue Representation," in Michael Marsh and Pippa Norris (eds.), *Representation in the European Parliament* [special issue of the EJPR 32: 2], pp. 165–84. Dodrecht: Kluwer.

Van der Brug, Wouter, and Cees van der Eijk (1999), "The Cognitive Basis of Voting," in Hermann Schmitt and Jacques Thomassen (eds.), *Political Representation and Legitimacy in the European Union*, pp. 129–160. Oxford: Oxford University Press.

Van der Eijk, Cees, and Mark Franklin (1991), "European Community Politics and Electoral Representation. Evidence from the 1989 European Election Study," in Hermann Schmitt and Renato Mannheimer (eds.) *The European Elections of the 1989 Parliament* [special issue of the EJPR 19: 1], pp. 105–28. Dordrecht: Kluwer.

Van der Eijk, Cees, and Mark Franklin, eds. (1996), *Choosing Europe? The European Electorate and National Politics in the Face of Union*. Ann Arbor: University of Michigan Press.

Wessels, Bernhard (1995), "Evaluations of the EC: Élite- or Mass-Driven?" in Oskar Niedermayer and Richard Sinnott (eds.), *Public Opinion and Internationalized Governance* [Beliefs in Government series, vol. 2], pp. 137–162. Oxford: Oxford University Press.

Zaller, John R. (1992), *The Nature and Origin of Mass Opinion*. Cambridge: Cambridge University Press.

CHAPTER 3

Identity and Electoral Participation: For a European Approach to European Elections[1]

André-Paul Frognier

Most comments on European elections make them out to be meaningless, or in any case deprived of any real meaning. One complains about the number of abstentions, considered an indicator of a deep lack of concern about European politics, which are continually on the increase despite an enlargement in European MPs' powers during the last two legislatures. As regards both participation and votes cast in favor of a particular party, the dependence of European electoral behavior on national behavior is stressed, and the term "second-order election"[2] is used to express this subordination. One is certainly forced to notice that the European voter does not inevitably vote in the same way as a voter does in national elections. European elections sometimes favor the opposition more than the ruling government. But voting is then interpreted as a "release," enabling the expression of dissatisfaction or the admission of new political actors onto the national scene. Between attraction for the vacuum and confusion as to what is at stake, the European voter appears as sort of faceless or with a funny nose, but certainly not as a ... European voter.

Every election must be approached as an individual event, and in a specific way, which does not mean in an *exclusive* way. In a state, for example, no one election is completely dependent on another, or completely independent. All elections in our countries are connected to one another. This applies to local elections and European elections alike. However, it is for the latter especially that the "dependency" aspect is constantly raised. The most often evoked cause is the number of abstentions. This reasoning is, however, valid only if these abstentions do not make reference to European political content. Furthermore, the "productive" effects of these elections—which can create new divisions and aggravate former or recent ones, or simply remind voters that divisions do exist—are rarely advanced. For example, the split with regard to Europe that has characterized French politics since the referendum on the ratification of the Treaty of Maastricht (1992) was brought to mind by the European elections of 1999.

We shall try to ask the question on the specificity of European elections from the point of view of electoral participation in these elections. We know that the percentages of participation vary greatly from country to country.[3] What do these differences signify? Is participation dependent on purely internal practices in states or is it motivated, at least partially, by European considerations? Let us consider the following hypothesis: Assuming that the European elections have a certain specificity, turnout will vary depending on the voters' identification with Europe. By "European identification," we mean the feeling one has of being European, certainly not exclusively—that would be unrealistic considering the level of European integration—but at least in a mixed way, as European and national. Countries with more voters identifying with Europe in this way should have a rate of participation higher than those where voters feel themselves less European. If this relationship were not verified, the field would be then more open for interpretations of the participation connected to the specificities of the internal political life of states.

This analysis will not definitively answer the question! A more exhaustive study should be situated at the same moment at the level of the individuals and of the states. A specialist in comparative politics will at once see the interest of carrying out such a study following the "micro-macro" procedure used by Przeworsky and Teune in their base work on comparative analysis.[4] Participation could at first be subject to a first analysis at individual level (micro) across all states—taken together and allowing an eventual discussion on the common determinants at stake. A second analysis would follow in which, in order to interpret the differences in explanation according to countries, one would bring in the institutional and historicocultural characteristics of states as "systemic variables" (macro). Such a study is not possible at the moment, because we still do not possess data stemming from any postelectoral survey carried out after the last European elections. This approach was already undertaken, at least partially, by Oppenhuis and by Franklin, van der Eijck, and Oppenhuis for the previous elections.[5] The present analysis is principally situated at systemic or macro level, and constitutes only part of a more synthetic future study; one, moreover, subject to the risks of "ecological fallacy."[6]

Measuring the European Identity

In order to study European identity at state level, we shall refer to the Eurobarometers, which have, for some time now, put forward a question supposed to measure this concept. The formulation of this question has changed repeatedly—more often than not in a relatively superficial manner—but in a more fundamental way in 1982 and in 1992.[7]

At first, from 1975 to 1979, there was only a question on feelings of belonging, whereby Europeans were asked what they felt they belonged to most: the world, Europe, their country, their region, or their municipality. From 1982 to 1992, the question concerned the frequency of those feelings. Did the subjects questioned feel themselves often, sometimes, or never "European citizens" or only "European" rather than citizens of their country? From 1992, the format was again modified: the question now being whether, in the near future, one will feel "only European," "European and national," "national and European," or "only national." However, another formulation was also used in 1989 and in 1994: People had to situate themselves on a scale of ten positions, of which one of the extremes (1) meant "exclusively national" and the other one (10) "at the same time national and European."

It turns out that Eurobarometers ask many other questions related to the question of identity. One knows, for example, that all the Eurobarometers have had, almost since the beginning, similar questions (called "trends" questions) enabling evolution to be measured. They concern attitudes with regard to European unification in general, to Europe's possible dissolution, to the fact that one's country is a member of Europe, and to the benefits accruing to a nation that belongs to Europe. However, it was already shown in a study on European identity based on feeling European or national that these questions did not really measure identity. They are more evaluative than identitarian because they do not sufficiently take into account the emotional content of identity. Therefore the two kinds of variables did not vary, over time, in the same way.[8] For these reasons, we are excluding "trends"-type questions from this study.

Table 3.1 displays some results illustrating the last three types of identity measurements we have just quoted. Table 3.2 shows the correlations among them. We added two trends variables to the second table: one relative to the question of whether the participation of the nation in Europe is considered a good thing, and one that relates to the benefits for a nation in belonging to Europe, in order to again test the specific meaning of the identitarian questions.

We shall use, in this analysis, the last formulation for measuring identity, used since 1992 (the one asking if, in the future, voters will feel only European rather than European and national, or just national and European,

Table 3.1. Measurements of identity, 1992–1998 (in %)

	Identity as European citizen (often/sometimes), 1992 (EB 37)	Euronational identity in the future, 1993 (EB 40)	European and national identity (scores from 6 to 10, 1994 (EB 41.1)	Euronational identity in the future, 1998 (EB 50)
Austria				51
Belgium	**51**	**65**	**57**	**56**
Denmark	**53**	50	40	48
Finland				**65**
France	**54**	**65**	**49**	**65**
Germany	37	56	**60**	54
Greece	**61**	**57**	36	50
Ireland	39	50	37	49
Italy	**57**	**70**	**55**	**71**
Luxembourg	**64**	**65**	**47**	**77**
Netherlands	43	**59**	46	**60**
Portugal	**64**	52	**48**	44
Spain	**58**	56	46	**66**
Sweden				40
Britain	33	37	26	38
Mean (*)	**51**	**57**	**46**	**56**

The results over the mean are in bold type.
*Unweighted.

or national). As this question was answered in 1993 and 1998, it has the advantage of covering two periods preceding European elections (1994 and 1999). We have dichotomized the question by distinguishing those who feel European and European and national—irrespective of the order—as opposed to those who feel national only. "Euronational identity" will be the name used to describe the identity based on the first of these two categories and that will vary from 0 to 100 percent.[9] It is at the same time, on the contrary, the measurement of a purely national identity (which is equal to 100 less the percentage of those professing a Euronational identity).

Reading table 3.1, one observes that although the European countries are different as regards this aspect of their political culture, the difference is a rather stable one. The same four countries—Luxembourg, Italy, France, and Belgium (in the order of 1998)—are above average on the four items of measurement, spread out over seven years. In 1998, the founder nations of the Common Market were not those who identified most with Europe, whereas this was almost entirely the case in 1993:[10] Spain and Finland appear in third and fourth positions. One notices, besides, that the averages are almost the same for the two comparable measurements of 1993 and 1998.

It is, consequently, not surprising to observe only a few changes in measurements in 1993 and 1998. One denotes significant progress only in Luxembourg and Spain. Decreasing feelings of identification are observed in Belgium, Greece, and Portugal. One will also notice that the inclusion in Europe of Austria, Finland, and Sweden has had no striking effect on the average perception of European identity. It also seems that the progress of European integration since 1993, especially as regards the new powers of the European Parliament, has had no influence on this average either.

Table 3.2, showing the correlations among the various variables used over the course of time, illustrates this relative stability. As one would expect, there is a stronger correlation (.837) between the two measures used in 1998 and in 1993, and formulated in the same terms, than between those of 1998 and 1992, when the questions were expressed differently. Besides, it confirms that trends questions do not belong to the same semantic universe, especially as regards the question on benefits.[11] It seems that the 1998 measurement of

Table 3.2   Correlations among variables of identity, 1992–1998

	Identity as European citizen: (often/ sometimes), 1992	Euronational identity in the future, 1993	European and national identity (scores from 6 to 10), 1994	**Euronational identity in the future, 1998**	*Participation in Europe: a good thing 1998*
Euronational identity in the future, 1993	.530				
N	12				
European and national identity (scores from 6 to 10), 1994	.231	.750			
N	12	12			
Euronational identity in the future, 1998	**.461**	**.837**	**.539**		
N	12	12	12		
Participation in Europe: a good thing, 1998	*.343*	*.365*	*.003*	*.492*	
N	12	12	12	15	
Benefit in belonging to Europe, 1998	*.365*	*−.008*	*− 333*	*.156*	*.866*
N	12	12	12	15	15

European identity can be taken as the reference variable for our study. It offers, indeed, sufficient guarantees regarding both validity and reliability.

Participation and European Identity

The preceding analysis rather contradicts results concerning the links among the different variables of support for European integration and electoral participation. For Franklin et al., participation does not seem to be connected with indicators of European attitudes.[12] On the other hand, Blondel et al., find links between electoral participation and some of the trends variables and other attitudes, such as satisfaction with the functioning of democracy in the Union, the approval of "the United States of Europe," etc.[13]

Figure 3.1 shows the relations between Euronational identity in the European states measured in 1998 (remember that this is the measurement closest to the last European elections) and participation in the elections of 1999.

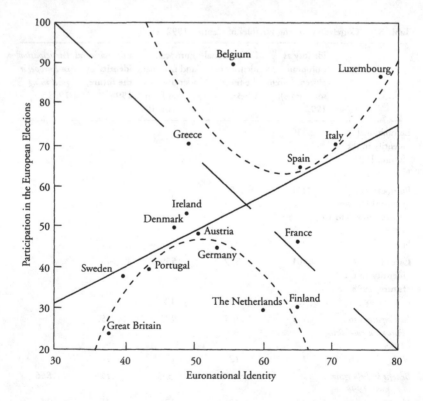

Figure 3.1 Euronational Identity and Participation in the European Elections of 1999, by Country

Our hypothesis of the positive relation between Euronational identity and participation is reasonably confirmed: It is expressed in the figure by the regression line. The best estimate is linear: The more Euronational citizens there are in a state, the higher the level of participation in European elections. Conversely, the greater the number of citizens feeling exclusively national in states, the lower the participation. The correlation between European identity and participation is .497.[14] Let us add that these relationships are specific and do not constitute a reflection of participation in national elections: Our indicator of identity is only slightly correlated to the participation in the last general election in every state (.213). Besides, correlations with the trends variables quoted in Table 2 appear significantly less clear: Coefficients are of .385 for the variable asking if the nation's involvement in Europe is a good thing and of .280 for that concerning the benefits to the nation of involvement in Europe. Turnout is associated more with identity than with the more evaluative variables generally used in this kind of analysis.

In figure 3.1, two groups of states are clearly separated (divided by the long dotted diagonal line). To the first—more identified in Europe and more participating—belong Luxembourg, Belgium, Italy, Spain, Greece, and France. In the second group—less identified and less participating—are Austria, Germany, Finland, the Netherlands, Portugal, Sweden, and Great Britain. However, one cannot be satisfied with the linear relationship observed. Indeed, the clouds of points on both sides of the diagonal do not display the same kind of linear relation. If there is a linear relationship between the two groups of countries, the situation is not the same inside these. It is certainly the case that a small number of points makes rather unpredictable the adaptation of a curve in the subgroups. Whenever there is a lack of other relevant data, this exercise will nevertheless be undertaken, consisting only in a preliminary recognition of the forms of the relations between the points.[15] There are some signs of curvilinear relationship, which are represented in the figure in finely dotted lines. For the most identified countries, the best estimate is that of a U-curve.[16] For the others, it is a reversed U.[17]

Whatever the interpretation of these curves, one finds inside each of the two groups relations among countries where a greater European identity entails more participation (with regard to the countries situated on that part of the curve rising toward the *right* of the figure) and also a tendency for, on the contrary, a greater national identity to lead to more participation (as regards countries placed on those parts of the curve rising toward the *left* of the figure). One can observe, when moving from Great Britain to Sweden, Denmark, Germany, and Austria, that one is following the direction of the general relation that associates participation with a greater identification with Europe. The same observation applies to relations among France, Spain, Italy, and Luxembourg. On the other hand, relations go in the opposite direction if one goes from Finland and the Netherlands to Germany, Austria, Denmark, and Sweden. Also, participation increases if one moves from Luxembourg and Italy toward Belgium, or from Spain and France toward Belgium and Greece.

These conclusions make the interpretation of figure 3.1 more complex: The linear relationship is partially a fallacious one. Data show that if the general orientation of the figure follows the sense of the hypothesis, one discovers that at the same time, the opposite relationships are also observed: Participation also increases with exclusively national identification! Anyway, these results reinforce the hypothesis of the "European" nature of participation (that is, the link between participation and the problem of European and national identities in Europe), as opposed to that of the internal specificities of states.

Structuring of the European States into Two Groups

The splitting of the states into two groups as depicted in figure 3.1, which is linked with identities, can be interpreted in various ways that are not mutually exclusive.

Institutional Aspects

The effect of these aspects is well known. In figure 3.1, countries with compulsory voting (Luxembourg, Belgium, Greece) occupy a very specific place and "pull" the figure upward. It is—obviously—a clearly discriminating factor. It is also the most determining institutional variable.[18] These countries do not, however, strongly influence the entire configuration: The correlation without countries with compulsory voting would remain at the same level (.506). Italy, however, poses a problem. On the one hand, Italy, since 1993, is no longer a country where voting is compulsory; on the other, it is likely that the voters have continued to turn out as if it were. In Italy, participation in European elections is almost at the same level as in Greece, but closer to that of Spain, where voting is not compulsory. If Italy is still considered a country with compulsory voting, correlation becomes much weaker and decreases to .315 because of the important position this country occupies in the linear structure of the cloud of points.

It is also in the countries of the same group that one observes the joint occurrence of European elections and national elections (Luxembourg and Belgium), European and regional elections (Belgium and Spain), and

European and local elections (Spain), which has to increase participation. Always in institutional terms, one will notice that voting took place on a weekday in several countries of the second group—Denmark, Ireland, the Netherlands, and Great Britain—although there is no major reason to believe that Sunday voting inevitably attracts more voters. If it is clear that the countries in this group have better scores on the scale of identity (Luxembourg and Italy are the two countries where Euronational identity is the strongest, whereas Belgium and Greece are close to the average), it is difficult, due to the lack of studies on this topic, to find a theoretical link between these aspects and the identity question. One suggestion is that the compulsory nature of voting favors a kind of identity resulting from the simple repetition of a behavioral practice.

Geographical and Cultural Aspects

If we leave the institutional ground, one can discover other aspects underlying figure 3.1 in large part. It is difficult, indeed, not to think of a north/south geocultural dimension, even though countries appear in each of the groups in disorder. This aspect is, however, almost perfect for 7 countries out of 12 (if one does not take into account countries where there is compulsory voting): One passes from Sweden to Denmark, then from Germany to Austria, approaching France, Spain, and Italy. Among these countries, the relation between identity and participation is almost perfectly linear. There is, however, an exception in this north/south classification: the position of Portugal. It is the only "geographical" exception in this two-group classification. A possible interpretation is of a kind of "Atlantic" identity, more open than the other Mediterranean countries toward Great Britain and the North of Europe. One will observe that these groupings of states do not refer to the dynamics on which the European Union was constructed.[19] Indeed, it does not align with the distinction between the founders of the Common market and the new members (the Netherlands and Germany are separated from the four others).

As far as this north/south divide is concerned, one can make reference to the vertical aspect of Rokkan's famous "conceptual map of Europe."[20] Accordingly, this north/south aspect can be interpreted as a religious aspect distinguishing the Protestant countries from the Catholic countries. However, there are no really coherent theories connecting the European identity aspects with electoral participation, although some will make reference to the so-called Vatican Europe. One could, however, think about the homology between involvement in a religion that is by nature supranational (from which nation-states had to emancipate themselves in the past) and attitudes in favor of a process of integration that also is supranational.[21]

Political Aspects

An interpretation other than the geocultural one could be found on data with which the practitioners of Eurobarometers are highly familiar. Asking repetitively a question on citizens' degree of satisfaction with the "way in which democracy works in one's own country" in each of the countries of the European Union, the surveys suggest that the countries of the north are generally more satisfied than those of the south. Table 3.3 shows that, with the exception of Luxembourg and Spain, the north/south divide is a rather clear one.

One could estimate that Euronational identity and the effect it has on participation in elections for the European Parliament appear as a sort of "compensation" with regard to this situation. The countries where satisfaction with regard to democracy is weak would give rise to a stronger European identity, Europe being considered as a possible factor of change. Figure 3.2 shows the relationship between satisfaction with democracy in one's own country and Euronational identity. One finds there, first of all, segmentation between the two groups of countries that we have associated up until now with north and south, with the exception of a light inversion between Finland and Greece (on both sides of the diagonal, the long dotted line).

But it is obvious that relations among these variables do not show many signs of a linear relationship. The correlation among them is only to the

Table 3.3 The position of the states on the scale of satisfaction with regard to democracy in one's own country, 1999 (EB 51)

States	% of satisfaction, 1999
1. Luxembourg	**83**
2. Denmark	**81**
3. Netherlands	**78**
4. Ireland	**74**
5. Spain	**71**
6. Finland	**67**
7. Germany	**66**
8. Sweden	65
9. ex aequo Britain and Austria	64
10. Greece	62
11. France	59
12. Portugal	57
13. Belgium	49
14. Italy	34
Mean (*)	65

The results over the mean are in bold type
* Unweighted.

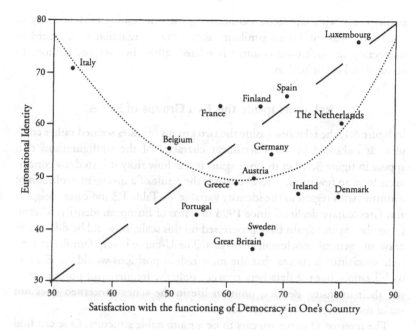

Figure 3.2. Euronational Identity according to the Degree of Satisfaction with Regard to Democracy in One's Own Country, by Country

order of −.033. Does "compensation" apply? In fact, it is necessary to leave—again—the domain of linear relations to see the existing relations among these variables. In a previous study, it had already been observed that relations between European identity and satisfaction with regard to democracy in one's own country had (on the part of the European citizen) a curvilinear character: Those who identified with Europe were situated at the same time among the most satisfied and the most dissatisfied.[22] One finds here something of the same characteristic at state level, as shown by the finely dotted line in figure 3.2, a figure mostly influenced by the extreme positions of Italy and Luxembourg, which are disentangled on the satisfaction variable.[23] Moreover, this curvilinearity is particularly evident if one takes into account only the most participating countries (from left to right of the figure: Italy, Belgium, Greece, Spain, Luxembourg). It consequently seems that European identity, depending on the country, acts as *compensation* (in Italy and Belgium especially) for those most dissatisfied with the functioning of democracy in their country, and as a *continuation* of an attitude of satisfaction in the other countries (like Luxembourg, Denmark, Netherlands, and Ireland). Besides, approximately the same results are found if one compares participation with satisfaction concerning the functioning of democracy in one's own country.[24] This similarity shows that satisfaction with regard to democracy in one's own country is related, albeit in complex fashion, to identity and participation.

Relations Inside the Two Groups of States

In figure 3.1, the relations inside the two groups of states seemed rather complex, as indicated by the curvilinear character of the configurations that appear in figure 3.1. Let us once again repeat how risky it is to draw conclusions from so few cases. The U-curve is the result of a divergent evolution of countries with regard to the identity variable. As Table 1.3 indicates, Belgium and Greece have declined since 1993 in terms of European identity, whereas Luxembourg and Spain have progressed on this scale. It would be difficult to draw any general conclusion from this. The U-curve is very familiar to attitude specialists: it means that the most radical positions would entail more mobilization, but the data here concern only the frequency of positions and not their intensity. Besides, political life in the states concerned does not make this interpretation a credible one.

The reversed U-curve appears to be a more stable structure. One can find it in relations involving the same types of variable with regard to the European elections of 1994. Identification with Europe has not greatly changed for the countries concerned; participation, however, has decreased in a rather constant way in these countries. One could find a more general interpretation if one observes that the top of this structure refers to an almost equal distribution of Euronational and exclusively national attitudes in states. One could consequently suppose that this situation produces in these countries more internal confrontations and, consequently, more electoral mobilization than in the lower-placed countries on the curve. One will notice the presence at the top of this curve of countries in which reside parties with strong anti-European consonance, such as Austria and Denmark. It also appears that other countries in which this type of party are present are also on this line, with a more or less equal share of the positions. This is the case with Belgium and Greece.[25]

The Politically Significant Character of Abstentions in European Elections

If participation is linked with European identity, abstention, which is the complement of it, must also, at least partly, be linked with it. One should therefore not expect abstention to be due only to purely accidental factors divested of political meaning as regards the very subject of the elections, i.e., Europe.

Blondel, Sinott, and Svensson conducted an in-depth analysis of abstentions during the 1994 elections.[26] The rate of abstention for the electorate was 43.2 percent. These authors showed, on the basis of the Eurobarometer that followed the election, that abstention was related to political attitudes.[27] The latter involve, in fact, a rather vague mixture of general attitudes of lack of interest and dissatisfaction toward politics as well as attitudes of opposition to Europe.[28] The authors call this "voluntary" abstention, as opposed to the "circumstantial" abstention that ensues from nonpolitical factors (absence, illness, etc.). What is the frequency of these voluntary abstentions? They amounted to 60 percent of abstentions during the European elections of 1994, which is equivalent to 25.9 percent of the electorate.[29] Among these, 24 or 10.4 percent of the electorate[30] at least would be inspired by motivations explicitly connected with Europe.

Blondel et al. also isolate the "Eurospecific voluntary abstainers," those that did not vote in the European elections for voluntary reasons although they did vote in the last national elections. One can estimate them at 40 percent of abstentions. For the electorate as a whole, that means 17.28 percent, suggesting that the figure of 10.4 percent for specifically European motivations is widely underestimated.[31] To those abstainers it is still possible to add a further 20 percent (8.64 percent of the electorate), covering "European and national voluntary abstainers," i.e. those who did not vote either in the European or in the national elections for voluntary reasons.[32] It makes sense to extend these results to the elections of 1999 (similar data are not available for 1999). Indeed, the nature of the abstentions can be considered as relatively stable over time, as can the abstentions themselves (even though they are slowly on the increase).[33]

The correlation—measured at state level—between abstention in 1999 and the percent of Eurospecific voluntary abstentions in 1994 is very high: .816. The relation is almost linear. In 1994, the correlation between the same variables was .810. Moreover, there is, on the other hand, no correlation with circumstantial abstainers ("Eurospecific circumstantial abstentions"[34]) either in 1999 (.041) or in 1994 (.012). These numbers indicate that abstentions in European elections are to a great extent associated with Eurospecific voluntary abstainers. This is additional proof of the political content of abstention, even though the specific European aspect of this behavior concerns only a fraction—albeit an important one—of these abstainers.

Conclusion: Identities Matter!

The analysis of the relationship, on European state level, between the European identity defined as "Euronational" and participation in European elections no longer makes the latter meaningless or dependent on national elections. This conclusion is valid only as far as electoral participation is concerned; the question of the relation between voting behavior for specific parties was not taken into account. In this perspective, the main result of the analysis is twofold: (1) The differences in electoral participation in states are connected to the extent to which the citizens of these states consider themselves European; (2) these links can be interpreted in several ways, even though our comments here are particularly exploratory ones, considering the lack of studies on the topic of identity in empirical European studies. We are faced with historicocultural processes relative to the territorial distribution of religious faiths and cultural areas, and also with aspects of the legitimacy of the states themselves.

The introduction of the identity variable seems to produce interesting results, again justifying the use of this all-too-often neglected variable—especially for the benefit of the trends variables—in European studies. The available data prevent more sophisticated analyses. However, these results seem sufficient to justify the analysis of European elections as "autonomous" elections without reducing them almost a priori to second-order elections. Even though national elections can influence European elections—and it is evident that this is the case to a certain extent (in the same way that national elections and local elections also influence each other)—the analysis of the latter justifies a specific approach. Also, this analysis encourages the study of European elections at a pan-European level in order to find out sources of variations that would not be apparent at national level.

Notes

1. Our database was compiled by Pierre Baudewyns and Mercédes Mateo-Diaz, whom we thank for that. We are grateful also to Annie Laurent and Bernard Dolez for the discussions we had on the subject of relations inside a "small N" population during a night session of the Congress of the AFSP in Rennes. We thank also Anne-Marie Aish for her advice.
2. The expression originates with Reif, K., Schmitt, H., "Nine Second-Order Elections: A Conceptual Framework," *European Journal of Political Research*, 8: 3–4, 1980.
3. See the chapter by P. Delwit in this volume.
4. Przeworsky, A., Teune, H., *The Logic of Comparative Social Enquiry*, New York: Wiley, 1977.
5. Oppenhuis, E., *Voting Behavior in Europe: A Comparative Analysis of Electoral Participation and Party Choice*, Het Spinhuis Publishers, Amsterdam, 1995; Franklin, M., van der Eijk, C., Oppenhuis, E., "The Institutional Context: Turnout," p. 306–322 in van der Eijk, C., Franklin, M., *Choosing Europe? The*

European Electorate and National Politics in the Face of the Union, Ann Arbor: University of Michigan Press, 1996. The approach is, in this latter case, more "macro-micro" than "micro-macro," as it begins with the macro level.

6. This error consists in applying inferences of the collective level to the individual level. See Przeworsky and Teune's, *op. cit.*, treatment of this topic.

7. The conceptual and theoretical content of these questions has never been clearly clarified. See on this subject Duchesne, S., Frognier, A-.P., "Is There a European Identity?" pp. 193–227 in Niedermayer, O., Sinott, R., *Public Opinion and Internationalized Governance*, Oxford University Press, 1995; and also Sheuer, A., "A Political Community?" pp. 25–47 in Schmitt, H., Thomassen, J., *Political Representation in the European Union*, Oxford University Press, 1999.

8. See Duchesne, S., Frognier, A.-P., *op. cit.*

9. The percentages are calculated by subtracting the "no answers" from the total of the samples.

10. It was certainly the case with the former FRG.

11. One will notice that the measures of 1992 and 1994 are only weakly correlated with the other ones. Using the measurement of 1994, Blondel, Sinott, and Svensson contested some results obtained by Duchesne and Frognier, with those of 1992 considered as inadequate (Blondel, J., Sinott, R., Svensson, P., *People and Parliament in the European Union. Participation, Democracy and Legitimacy*: Oxford, Clarendon Press, 1998, pp. 63–65). In fact, the two types of measurement do not have the same meaning: The 1992 measurement concerns the frequency of European feeling, whereas the 1994 measurement concerns the intensity of this feeling. Let us note that, as they are both correlated to the other two identitarian variables, both are pertinent, each in its own way.

12. Franklin, M., et al., *op. cit*, p. 322. For them, studies showing links among these variables concern intentions of participation, while these relations disappear during real votes. Cf. on this subject van der Eijk, C., and Schmitt, H., "The Role of Eurobarometer in the Study of European Elections and the Comparative Development of Electoral Research," in Reif, K., and Inglehart, R., *Eurobarometer: The Dynamics of European Public Opinion*, London: Macmillan, 1991, pp. 245–257.

13. Blondel, J., et al., *op. cit*, p. 83.

14. It takes into account 24.7 percent (sign.: 0.06) of the variance. Although we are not working on a sample, the measurement of significance can be interpreted as an indicator of specificity of the curve: The smaller it is, the greater the chance that the curve is specific for the data. One finds a similar correlation (.498) if one compares European elections and the last national elections for a difference in participation. The figure that results from it is also very similar. Only Ireland passes from one group to the other.

15. It is evident that the small number of cases makes any measurements highly dependent on the presence or the absence of each case.

16. The curve comes from a quadratic regression by the least squares method. The adjustment was made for all points in the group. However, it becomes a valid estimate only if one considers the cases of Belgium, France, Spain, Italy, and

Luxembourg, without a real difference in the drawing of the curve. It then takes into account 80.3 percent of the variance (sign.: .197).

17. Here, the quadratic estimation fits almost perfectly all the data, with 78.9 percent of the variance explained (sign.: .009).

18. One finds the same result in other studies, e.g., Franklin, van der Eijk, and Oppenhuis, *op. cit*, pp. 322–323. The authors observe the result in a bivariated analysis as in a multivariated one. Compulsory voting also absorbs the effect of individual-level variables like interest in politics.

19. We have already noticed the absence of relations between the time of joining the Union and European identity. See Duchesne, S., Frognier, A.-P., *op. cit.*, 1995, p. 223.

20. Rokkan, S., *Cities, States and Nations: A Dimensional Model for the Study of Contrasts in Development*, in Eisenstadt, S. N., Rokkan, S., eds., *Building States and Nations*, Beverly Hills: Sage, 1972.

21. Oppenhuis shows that in 1994, religious practice was, after age, the sociodemographic variable most associated with participation. *Op cit.*, p. 80.

22. See Duchesne, S., Frognier, A.-P., *op. cit.*, p. 217.

23. The quadratic regression takes into account 30.5 percent of the variance for figure 3.2.

24. Correlation is of −.224 and separates about the same countries' groups; here also, the relation shows signs of curvilinearity, although less obvious than in the previous case.

25. But it is certainly not the case for Ireland.

26. *Op. cit.* see particularly pages 40–54.

27. Eurobarometer 41.1.

28. Data were generated from answers to an open question in which the absence of explicit references to Europe did not inevitably mean that it was not on people's minds. Percentages of the motivations of the abstentions were as follows: lack of interest: 43 percent; lack of confidence or political dissatisfication: 27 percent; lack of knowledge: 24 percent; vote has no effect on the functioning of institutions: 11 percent; opposition to the European Union: 8 percent; dissatisfaction with the electoral system of the European Parliament: 8 percent; dissatisfaction with the European Parliament as an institution: 4 percent; vote rarely or never: 4 percent; European Union viewed as ineffective or nonpertinent: 4 percent.

29. Cf. Blondel, J., et al., *op. cit.* p. 42. It is a projection of the results obtained in the Eurobarometer for the complete electorate.

30. According to the same calculation as in the previous note.

31. One can presume that those who do not go to vote in the European elections but have voted in national elections, while expressing political reasons for their abstention, do not behave without some European preoccupation.

32. 8.6 percent of the electorate, always according to the same calculation as before.

33. The correlation between the participation in the European elections of 1994 and those of 1999 is .828.

34. Cf. Blondel, J., et al., *op. cit.* 1998, pp. 42 and 167.

CHAPTER 4

The Faltering Europeanization of National Campaigns

*Jacques Gerstlé, Holli A. Semetko,
Klaus Schoenbach, and Marina Villa*

In June 1999, European Parliamentary elections were held for the fifth time in 20 years. One would have thought that after two decades of voting experience at the European level, voter acceptance for the institution would take hold in the countries that had participated in the European construction since its very inception. We now know that this did not happen: Electoral turnout has never been so low despite a priori favorable conditions such as the well-publicized inauguration of the euro zone or the implementation of the Amsterdam Treaty. The current situation is paradoxical in that the European Parliament is in a period of expanding power following the Maastricht and Amsterdam Treaties and the demission of the European Commission soon after the March 1999 crisis; moreover, it is a unique European institution, having its legitimacy based on universal suffrage. However, the Parliament has only been able to provoke a weakening voter turnout.

To understand the reasons for this defection, we compared national campaigns for the election of the members of the European Parliament in Germany, France, Italy, and the Netherlands. All four of these countries are founding member states of the European Union, for which they equally lend their support, and they also have a common turnout record with respect to the June 1999 elections. Electoral turnout does not necessarily imply strong EU support, since a vote can be cast for an anti-European platform. However, it is still surprising to note that even in a country such as the Netherlands, there exists an important gap between low voter turnout

(30 percent of the registered voters) and widespread pro-European feeling. Although the Dutch vigorously support the EU, their participation in European elections is consistently low (on average 47.8 percent for the period 1979–1994). The wide gap between public support and voter turnout increased dramatically in 1999. In these countries very much in favor of EU membership, there are structural causes for low voter turnout, which are aggravated by local conditions under which elections are held. By focusing on the latter, we may understand more fully the gap and be able to differentiate between immediate or local and structural variables. Is declining turnout rate attributed to: the weakening of public support for the European ideal? the failing legitimacy of European institutions in light of the Santer Commission crisis? the strengthening of anti-European electoral debate? the Kosovo aftermath? or perhaps specific political conditions at the national level? Can an analysis of the electoral campaigns and of the circumstances under which they were held explain the all-time low abstention rate in European elections, and whether or not it is part of a long-term trend of low voter turnout dating back to 1979? To that end, it would be useful to carefully define the different levels of attitudes and behaviors to facilitate a comparison among the four countries and one between them and the EU (or the EC) (see table 4.1).

How has national opinion regarding the EU evolved since 1994? The Eurobarometer surveys sponsored by the European Commission can shed some light in this respect. The Dutch and Italian support rates are very much above the European average. Between 1994 and 1999, public opinion in favor of the EU in the Netherlands fluctuated around 70 percent, reaching a high of 73 percent in April of 1999. Since 1994, Italian support has ranked consistently above 60 percent and eventually leveled off at 62 percent in April 1999. French support for the EU follows the European pattern, which fluctuates around an average rate of support of 50 percent and has settled at 47 percent according to the most recent polls. Finally, the Germans have been well under

Table 4.1 National opinion regarding the EU (1994–1999)

	Germany (%)	France (%)	Italy (%)	The Netherlands (%)	EU (%)
Support for the EU, April 1999	44	47	62	73	49
Electoral turnout June 1999	45.2	46.7	70.8	29.9	49.6
Decrease in turnout, 1994/1999	14.8	6	4	5.8	7.2
Average turnout, 1979/1994	61.2	54.7	81	47.8	59.7

the European average, with a support rate edging toward 40 percent and attaining 44 percent in 1999. Be that as it may, the common denominator of the four case studies is the significant drop in national support between the fall of 1998 and the spring of 1999. It seems reasonable to connect the drop in support, at least in part, to the crisis and retirement of the European Commission on March 15, 1999. Our aim here, however, is not to evaluate the campaign's impact on the distribution of votes, nor to assess the new balance of forces, whether that be in the individual countries or within the European Parliament. Given space constraints, our focus will be limited to the reconstitution of the environment in which opinions were formed and in which elections were held as well as efforts made by people to influence that environment. To that end, we begin by comparing the degree and form of campaign visibility in the four countries. Then we identify the content of electoral messages, examining the extent to which they included specifically European considerations, and their eclipse. Finally, our study concludes with a brief analysis of the perceptions of the campaign and of electoral mobilization.

The Limited Visibility of the Campaigns

The campaigns in the four European countries were distinguished by two phenomena having adverse political consequences. Not only was platform communication poor, but its visibility was also reduced by prominent international issues (the Kosovo crisis and war) and by overbooked political agendas on the national fronts. This limited visibility resulted in an overall weak mobilization of the electorate.

In Germany, the campaign was low-key, very modest, unapparent. Contrary to previous elections (Semetko, Schoenbach, 1999), the campaign was carried out only two (SPD) or three (CDU) weeks preceding the vote and was consequently at variance with the other campaigns, which lasted nearly twice that amount of time. German political parties spent very little on the campaigns. The CDU disbursed 10 million euros (including 7.5 million for advertising), that is, nearly four times less than what had been expended during the federal elections held nine months earlier. The CSU paid out 3 million euros, the FDP 1 million euros, and the Grünen 250,000 euros. Particularly instructive is the amount of money spent on advertising by the main party of the government coalition. The SPD allocated a mere 5.5 million euros to acquire a 90-second TV spot, radio time, a spot shown in movie theatres of the major cities, six billboard advertisements, and newspaper coverage (in *Der Spiegel* and *Bild am Sonntag*). The TV spot was diffused for free by public channels and cut down to 45 seconds by private channels to allow for distribution fees. The radio spot was broadcasted solely on the main frequencies and appeared seven times during the whole campaign. Such limited advertising efforts were hardly compensated for by any personal involvement on the part of politicians, as indicated by the scarcity of individual and common meetings. Wolfgang Schaeuble, leader of the Christian Democrats, was the most active politician, making 40 public speeches, whereas Joschka Fischer, minister of foreign affairs—who is also the most popular politician in Germany—participated in but one of the Grünen's meetings.

The information media did not grant considerable attention to the elections. They obliged the national public channel, ARD, and broadcasted five electoral programs during the week preceding the vote. In fact, these programs did not exceed 15 minutes and were shown after 11 P.M. The other national public channel, ZDF, aired one program devoted to the elections. In view of this deficient electoral communication, the German government allocated funds to set up a small campaign aimed at raising voter turnout to a decent level.

In the Netherlands, despite considerable efforts by political parties and the media, the campaign was barely visible. Officially, the campaign started on May 29 for most of the parties involved, even though in March local elections were held as well as the elections for the upper house, which thus provided an opportunity to introduce a European perspective into the campaign. Since 1994, the Social Democratic leader, Wim Kok, has headed a government coalition comprising his own party, the PvdA; the VVD; and the D66. During the campaign, the coalition government spent approximately 400,000 euros, essentially on advertising space (TV, billboards, booklets, newspaper coverage). From the outset to the close of the campaign, the motto was "Wim Kok votes Max van den Berg" (i.e., the prime minister votes for the leader of the Social Democratic ticket), which attests not only to the national color of the electoral debate, but also to its proclivity to personalization. The center-right party, VVD, equally resorted to marketing to better determine the preferred target of decided voters (2.1 million), the majority of whom were between twenty and fifty years old and highly educated. The VVD dedicated the greatest amount of its budget (228,000 euros) on advertising space for TV, radio, and the printed press. The D66 center-reformers became involved in a political quagmire, which ruined their campaign just prior to the European elections. They were the subject of one of the three national crises (the governmental coalition crisis, the dioxin issue, and the Bijlmer scandal) that overshadowed the campaign. To be sure, the free negative publicity canceled and even overrode the effect attained by D66's campaign expenditures (46,000 euros). The main opposition party, the CDA's Christian Democrats, relied on a famous and popular ticket leader, Hanja Maij-Weggen. The former Parliament member was well known for her feminine and family-oriented programs, which brought a very personal tone to politics. The CDA's TV and radio advertising spots stressed local politics, that is to say the level at which candidates organized meetings under the patronage of Parliament members. The expenses of the CDA climbed to 137,000 euros. The left-wing environmentalists, the Groenlinks, who were also part of the opposition, construed their campaign to display their organization's professionalism. They allocated 183,000 euros on an array of publicity forms: three-minute national TV spots, 30-second local TV spots, national press, two opinion magazines, booklets and pamphlets. The Groenlinks organized only three meetings to present the party's platform and to attract media attention. The alliance of protestant parties devoted most of its budget (110,000 euros) to advertising and to arrange 12 meetings. It is crucial to note that the alliance decided not to include their national leaders in the Europe campaign, fearing it would undermine their credibility. Lastly, it is surprising to observe that the radical-left party, the SP, spent more than 228,000 euros on political advertising to protest against government policy and European integration. With such a message, they entered into the European Parliament, where they won three seats.

The media coverage of the campaign in Holland was almost nonexistent according to findings taken from a study of the four main daily newspapers and the public channel NOS, which has the greatest number of viewers. Indeed, between May 29 and June 10 (the voting day), TV news covered two subjects related to the elections, the first concerning voter intentions to abstain and the second involving jobs created by European financing. In the printed press, daily newspapers published between two and eight articles during the last 12 days of the campaign. In general, published information was on personal details of ticket leaders and the fear of a strong abstention rate grounded in Euroapathy. There were also articles on the PvdA's campaign financing, which echoed the corruption problem in European politics raised by certain parties and which also made reference to the Santer Commission crisis in Brussels. The various standpoints of the political parties were hardly mentioned, except when common positions were established.

In Italy, the campaign was intermittent and often obfuscated. To be sure, the national political agenda was so packed with issues that alone it could have fully absorbed the public's attention. Moreover, in Italy, as elsewhere, the Kosovo war completely captured the public interest, thus rendering secondary the topic of European elections. In the Italian campaign, it was difficult to single out a specific time period during which a continuous development occurred. This cannot be explained by a rise or fall in public interest but rather a general lack of concern, which was interrupted by a temporary focus on European issues. For example, it is worth recalling that two times during the campaign, tremendous visibility was attained. The first occurred during the ticket presentation period, which ran from the first week in April through the first week in May. The race to gather signatures and the makeup of the ticket were subjects of interest for the media, which could focus on the competitive and personal aspects related to these activities. The concomitant presidential elections allowed for the exploitation of another political arena to campaign in, as in the case of Emma Bonino. The European campaign benefited from relative visibility for the second time during the two weeks preceding the vote, particularly as chances increased for a military agreement in Kosovo. During this brief period, political parties were able at least to find an audience even if they never really set up genuine European platforms.

The advertising campaign in daily newspapers, on TV, and on billboards was expensive and prompted polemics on the exact costs and financing of the campaigns. Expenditures on spots and communication activity were particularly high for Forza Italia and the Bonino teams. The Radical Party had to regularly justify its budget in the media and claimed that they were ready to sell the party's radio station to pay for their campaign expenses. The use of the Internet rose dramatically at various times to promote either the tickets or the candidates. In the field, the amount of organized meetings and events was quite low, with the exception of Prodi's Democratic "Asinello" train, which traveled throughout the country promoting the party platform. One of the reasons why politicians relied extensively on advertising and the Internet was to compete with the overwhelming level of international (Kosovo) and national (other elections) political news, which prevented the public from accessing other issues. To get a campaign message across, one needed to make a great deal of noise in a short amount of time or use more unconventional tools of information media. A main feature of the campaign was a high level of personalization, which influenced both communication activities controlled by political actors and the way the media treated information. For example, journalists from Telegiornale 1 and RAI stressed the weakness of discussion on European campaign issues. This corresponded to a classical form of interaction between the media and the politicians, according to which they blamed each other for the irrelevance of the electoral debate. Yet, the media, which tended toward information personalization, incited a preference for national politics over European politics in accounting for the electoral situation. The trend implies a double framing effect, that is, European issues have to give way to national ones, and political competition prevails over public policy agendas.

The irregularity and poor visibility of the French campaign made it similar to its Italian counterpart. These features were found in the communication activities controlled by political parties and in the media coverage of the campaign. Unlike the other three case studies, French rules applying to electoral communication are very constraining. Electronic media advertising and posters, which are the essential means of communication elsewhere, are strictly banned in France. Consequently, it is necessary to rely on classical means of communication, therefore implying a focus on the composition of tickets and on personalization, the formation of support committees, and party-member and local activities (e.g., meetings, public assemblies, demonstrations), as well as a reliance on the way the media reckons to cover the campaign. As a result, every main ticket is composed of prominent personalities who are chosen for their degree of notoriety. As an alternative to the prevalent mass-media communication tools, meetings were the most frequently used form of communication. From April 12 until May 31, 95 meetings were held with members of Jospin's government and 65 were planned for the first 12 days in June. The most dramatic meetings called together government leaders and Socialist secretaries in France and abroad. They aimed to show how well coordinated the Socialist government was throughout the EU (e.g., Milan, March 1). The EPP was less successful when trying to plan such spectacular international meetings, an example being the meeting of May 30. Les Verts equally attempted to gather support from European environmentalists by staging a meeting in Paris. Although some of the initiatives taken by the parties were unsuccessful, they all were sincere efforts to Europeanize campaign activities through different group formations. Within nine months, Les Verts held 400 meetings, 45 in which Daniel Cohn-Bendit participated. The Communists claim they organized 1,400 "public initiatives" during the campaign, and had the largest budget allocated to the campaign, with 6 million euros, whereas the Socialists invested 5.6 million euros. Moreover, each main party created a Web site to provide easily accessible information on the electoral party list, the platform, the forthcoming communication activity, meetings, and leaders' and candidates' speeches, and in some cases, to present an analysis of current issues.

In addition, the campaign included televised debates. For example, on Sunday, May 30, the program *Public* on channel TFI staged a debate between François Hollande and Nicolas Sarkozy, which reached an audience of only 2.1 million viewers; it was followed by a debate between François Bayrou and Cohn-Bendit on June 6. On June 7, a debate reminiscent of the 1994 program brought together eight ticket leaders on France 2 that created the same overall feeling of cacophony for the electoral message. The official campaign on public TV and radio stations spread across two weeks, from Monday, May 31, to Friday, June 4; and then, from Monday, June 7 through Friday, June 11. The tickets of the RPR-DL, the UDF, the PS, and the PCF were each allocated 30-minute speech slots, or as much time as the whole period granted to the other 16 tickets combined. Since the speeches were aired during peak audience viewing and were rediffused, these public channels reached daily approximately 6.9 million individuals during the ten days of the official campaign.

Using official statements produced by the Commission Supérieure de l'Audiovisuel (the French FCC), we can evaluate the campaign visibility in daily televised news broadcasts. In reality, the May–June 1999 campaign visibility in French television news was twice as weak as that during the last European campaign held in 1994, and five times weaker than the legislative elections in May–June 1994. Focusing on the time period devoted to the electoral campaign by the five national channels broadcasting TV news, there is evidence of an intensification of the coverage during the last month before the vote, as opposed to the preceding 42 days. This period represented not only the period during which all the political actors made their speeches, but also the time when political groups were formed and when presentations, commentaries, and quotations were made. For the nine main tickets supported by political parties, visibility was doubled because they were granted additional TV news coverage (from 8 minutes to 18). Certainly not every ticket experienced to the same degree this upward trend. The most dramatic ascent was for Les Verts, whose ticket went from 3 to 76 minutes. The PS visibility increased from 52 to 156 minutes, and Pasqua platform's visibility went from 24 to 66 minutes. As for the visibility of the RPR and UDF, the only ones to trend downward, it decreased from 136 to 62 minutes and from 79 to 62 minutes, respectively. This illustrates the degree to which the Pasqua ticket caused a media prejudice against the RPR as resources were divided. If we then take into consideration the pro- or anti-European positions, it is clear that the most pro-European tickets (Les Verts, UDF, PS) dominated daily news coverage, with approximately five hours of airtime, as opposed to anti-European tickets (Pasqua, the sovereignists, the radical right), who appeared but for two hours and 22 minutes. In this respect, TV news coverage of the electoral campaign contributed to the development of a very pro-European integration mood.

The campaign visibility by printed press coverage was assessed by counting the number of articles in the four major daily newspapers: *Le Monde*, *Libération*, *Le Figaro*, and *Les Echos* during the six weeks spanning May 3 through June 13. As in the case of TV news, we observe greater visibility of the campaign in its final phase, since the last three weeks (half of the period studied) provided 58 percent of the published articles. Quantitatively, the content of the articles was overwhelmingly tied to national considerations of the campaign. In the final two weeks, the written press broadened its scope by covering electoral campaigns in other countries of the EU. Nearly 90 percent of the articles dedicated to the elections were published in the last two weeks of the campaign. Similarly, 70 percent of the articles that focused on Europe, its construction, its institutions, and its affairs were published in the final stage of the campaign. The Europeanization of the campaign's media coverage developed in the last part of the two weeks preceding the ballot. Europeanization in this context signified an analysis of the campaigns outside France as well as an analysis of European issues. On this point, and this can only be expected, the written press became more tolerant toward increasing coverage, as compared with television, the latter being more constrained by airtime availability.

The Electoral Message and its Eclipse

We will examine two points of each national campaign. On the one hand, to what extent were European affairs a core issue in the structuring of the electoral message? On the other hand, which factors contributed to their eclipse?

In Germany, European elections took place nine months after Gerhard Schroeder's party (SPD) came to power. The political debate did not place European construction at the top of its agenda. It was only the Grünen—and they were to lose five seats—who tried to lead a genuine European Union; their motto was "European without hesitation." The essence of their message was to view the European Union as a means to solve different kinds of problems, such as those affecting the environment, women's rights and employment, and to address issues related to regional identity. The SPD slogans effectively revealed the crux of their electoral message: The saying "The SPD, good for you, good for Europe" claimed that European and domestic politics were intertwined and that the Social Democrats took care of German national interests within the EU framework. When they said "Our European policies do not target Brussels, but Bochum, Chemnitz, Bamberg, and Kiel," the aim was to underline how much the party was concerned with the concrete repercussions and local effects of European activities. In the slogan "Europe needs fair decisions," the party brought up ongoing complaints about the German financial contributions to the EU. During the last weeks of the campaign, the message became increasingly insistent on the chancellor's personality, since the SPD was obviously trying to take advantage of his presumed popularity to buoy up its own drastically weakened popularity, which had lost seven points, dropping to a 35 percent rate of confidence. There was also evidence in the CDU's strategy of trying to stimulate a sanction vote: the slogans "The one who fails in Bonn has to give up Europe to the others," and similarly "Europe must be build correctly," alluded to the issues of crime prevention, the politics of the elderly, the strengthening of democracy, the improvement of education, the Euro, employment, and taxes. Undoubtedly, the CDU attempted to convince people to view the

European elections as an opportunity to teach the Social Democrat/Grünen coalition a lesson. The final results proved they were right to adopt such a strategy. The CSU laid the stress on the personality of the Bavarian prime minister, Edmund Stroiber, and the notion of strength, as seen in their expressions "Moving force" and "A strong Bavaria for Europe." The negative aspects of their campaign consisted of discrediting the governmental coalition's handling of public health and tax policies. The FDP's electoral message focused more on domestic issues, as they urged people to "express [their] dissatisfaction with the Greens and the Reds"; the question on a billboard that showed a naked posterior read: "How do you feel after nine months of Red-Green?" suggesting that voters "got taken," which is the translation of a rude expression in German, *verarscht*. The former Communist Party, the PDS, directed their criticisms of the EU on the basis of their refusal to support the Euro currency and their stance against military intervention in Kosovo ("Building Europe without resorting to weapons"). As for European policy, the platforms of the political groupings favored a general discourse rooted in principles such as transparency and voter turnout instead of taking definite positions on public policy issues. The overshadowing of the debate on a European policy was a result of news being overwhelmingly dedicated to the Kosovo crisis and, to a lesser extent, to domestic matters like the implementation of a budget adjustment plan of 30 million DM for 2000.

In the Netherlands, the strong European feeling explained why no one expected European issues to cause party divisions. For instance, members of the Purple Coalition (center left) and the CDA Christian Democrats were in favor of a European constitution. And the Greens (Groenlinks), who described themselves as "critical pro-European," voted nevertheless against both the Amsterdam Treaty and the euro. The protestor parties' ticket was an exception in that it took a stand against Europe. In Strasbourg, the three European Parliament members of this anti-European rightwing entity sat with the group Europe of Democracies and Diversities and the French Chasse, Pêche, Nature et Tradition (Hunting, Fishing, Nature and Tradition). The most recurrent theme of the electoral debate was the struggle against fraud and corruption in European institutions. This widely shared concern, as well as the Santer Commission crisis, most likely failed to persuade the indecisive members of the constituency to participate in the elections. The PvdA campaigned with the slogan "A strong and social Holland within a strong and social Europe." Employment, security, and environment could only improve in the context of the EU. The accent was placed on benefits that the Dutch would receive and, therefore, on what the local stakes were. The VVD's ten-point platform argued that questions henceforth need to be addressed as both

Dutch and European in nature. As for D66, the third coalition partner, it developed five main themes: foreign policy and a common defense policy (structured around the Kosovo case); democracy and integrity, required to instill openness in Europe; social and environmental affairs as European problems; the eastward expansion of Europe; and EU political asylum and refugee policies. Christian Democrats adopted the slogan "Strong locally, at home in Europe" in promoting a democratic, united, secure, and social model for Europe. The Groenlinks, who went from one to four European Parliament members, used the motto "Schoon en eerlijk" ("Beautiful and honest"): i.e., environmental issues have no boundaries and honesty must inspire the functioning of European institutions and underpin their social and immigration policies. Protestor parties upheld a Europe built on cooperation but not a federal Europe, which explains why they wanted to retain national control over education and public health. The other ticket opposed to European integration was the extreme left-wing SP, who defined themselves as a protestor party ("Vote against, vote SP"). The SP cultivated a critical view of the EU and preferred to develop a nationally based campaign on the presumption that Europe was too far removed from the Dutch.

The overall electoral debate did not reach the public, whose attention was diverted to three national political crises, the most important of which involved the government coalition and began in March 1999 after the legislative elections were held. The coalition's founding agreement included a constitutional amendment relative to referendums. Yet, a VVD senator refused to vote in favor of it, which meant that qualified majority was not attained. For two weeks thereafter, every possible political-media aspect of the governmental crisis was elaborated and prognoses made on the future organization of the new elections. The second crisis originated in Belgium and involved the dioxin scandal. The minister of agriculture, a VVD member, did not immediately discuss the threat of dioxin when he became aware of it, but waited a few weeks. The third crisis was linked to the Bijlmer ordeal, the parliamentary investigation into what the government knew about the El Al airplane crash of 1992 in poor Amsterdam suburbs, and the hypothetical effects caused by the freight of weapons and chemicals. Parliamentary hearings were televised on a daily basis in April 1999, and people expected to discover how certain secretaries covered up the inquiries that would have otherwise led to the resignation of the government. The coalition and dioxin crises more or less displaced the Bijlmer scandal from the agenda. Briefly formulated, the national agenda in the Netherlands had the effect of reducing the visibility of the European campaign to a bare minimum.

In Italy, domestic issues also prevailed and eclipsed the European debate. The problems stemmed from an electoral "glut" and from the fact that

European elections were internally exploited. European issues were not, in fact, a source of division in the electoral debate, since Euroskepticism was marginal and confined to the radical right. Right- and left-wing parties supported a united Europe; although the Communists were against Maastricht, they were nevertheless in favor of the euro currency and held a rather subtle position toward the question of intervention in Kosovo. Even the Lega Nord separatists backed European integration as long as it was patterned after a regional model. In the end, the electoral debate was not sufficiently creative to gather public attention, which was highly solicited elsewhere. As regards the Bonino ticket, of which the "head" was launched by the presidential campaign, the former European commissar carried great credibility. The Democratic ticket led by the former council president Romano Prodi was bolstered by the leader's new legitimacy, since on May 5 he was to assume the office of the presidency of the European Commission. Finally, Fini's National Alliance and Segni's movement were in agreement on European issues.

The European vote was sidetracked by two previous elections, namely, the presidential election and a referendum on changes in the electoral system. President Scalfaro's mandate was to come to an end on May 28, and to avoid a superposition of elections, President Ciampi's election based on indirect suffrage took place on May 13. On the very same day he was sworn in, President Ciampi reiterated his concerns about European integration. Another conjunction of elections for the European campaign came from Emma Bonino, member of the Radical Party, whose popularity grew from her humanitarian activities as a European commissar, notably in Kosovo. On television, Bonino launched her communication campaign on April 7 for the presidential elections, most likely anticipating its effects on the European elections. The slogan "Emma for President" was indicative of a genuine presidential campaign—unique for a political system in which presidents are not elected according to direct universal suffrage. All of the publicity instruments were being utilized to promote Bonino's candidacy—TV spots, billboards, mailings—and there is no doubt that this personalization process helped Bonino's ticket a month later. The second electoral episode dealt with a referendum on reform of the electoral system designed to reinforce the majority principle. The vote occurred on April 18, with the campaign beginning approximately one month beforehand. That made three elections organized in Italy between mid-March and mid-June, in addition to municipal and provincial elections, which were held on the same day as the European vote (June 13). In three months, a multitude of electoral agendas were prepared, superposed, and intertwined in a competition that led to the overshadowing of the weakest issues, that is, the European agenda. The parties launched their campaigns late and were to acquire strong visibility only in the final phase. The political right perceived the elections as a litmus test for majority legitimacy and underscored their habitual themes. Similarly, the left highlighted the social dimension of Europe and the possibility of a "Third Way," following the declaration made conjointly by Blair and Schroeder. To her dissatisfaction, Emma Bonino was the only candidate to focus on European questions, which coincided with her previous experience and reputation, just as her humanitarian causes and position on a common security policy spurred her to hope for a "European United States having a single foreign policy instead of fifteen." There was also a bloody event that affected the electoral debate: Massimo D'Antona, staff member of Minister of Labor Bassolino, was assassinated by the Red Brigades. D'Antona was responsible for the elaboration of the government labor project, which was to be presented for approval by the European institutions. This dramatic event brought issues related to unemployment and labor policy to center stage. The second reason why European issues were clouded was that European elections became an internal device to ascertain the balance of political forces. At the heart of the campaign stood Silvio Berlusconi, who gave clear instructions to interpret the European elections in this manner. Berlusconi reckoned that if the majority parties failed to get 40 percent of the votes, anticipated elections would have to be arranged. That meant, in essence, turning an election designed to select members of the European Parliament into a test of majority legitimacy.

Be that as it may, the poor media coverage of the campaign and the personalization of information did not imply that Europe was absent from the nonelectoral news, but rather was considered the source of a possible priming effect. Everywhere coverage on the European Union, as distinct from information on the Italian campaign, fueled the news as it was defined by the media. As elsewhere in Europe, the media focused narrowly on the Kosovo crisis, which profoundly affected the public's attention. The primary effect of the Kosovo crisis was undeniably an overshadowing of the European elections. The secondary effects of the Kosovo crisis played into Italian politics. During the first phase of the campaign, news on the war was frequently directed at understanding its impact on Italian politics: the Italian role in the EU, the likelihood of a common security policy, the refugee problem, EU eastward expansion, European political parties and their view on the military intervention. During the final phase of the campaign, at the fore were Italian peace proposals, the reactions they triggered, and the role of Italian politicians in the Union and in the peace process. To a lesser extent, the EU institutional agenda was the subject of media interest because of the Santer Commission crisis and the nomination of Romano Prodi as leader of the new Commission. This news carried even greater weight given Prodi's candidacy on the Democratic ticket, and conversely, the outcome of Prodi's bid for the presidency of the Commission heightened the visibility and legitimacy of his candidacy and that of the Democratic ticket. In European affairs beyond the Italian borders, three types of problems bore upon the electoral debate. Firstly, the dioxin-contaminated meat in Belgium raised concern over health and security problems and the politics of common agricultural issues. Secondly, electoral campaigns abroad were granted coverage on two specific occasions: first during the European Socialist meetings in Milan on March 1, at which a common manifesto was adopted, and second, when Blair and Schroeder made their joint declaration, which appeared to diverge from the previously mentioned manifesto and to isolate the French Socialists' orientation. Thirdly, the closing of the campaign in Europe was characterized by the rising threat of a very low voter turnout, which would confirm citizens' lack of interest in European matters. This feeling was widespread in Italy, where the May referendum ended in failure because the quorum (50 percent turnout) could not be achieved.

In France, European issues were apparently much more central to constructing the electoral debate. Nevertheless, Europe was overshadowed by the way in which the political-media agenda construed the news. Firstly, political positions on European integration were structured within the federalists-sovereignists opposition. On the federalist side were the UDF party, which claimed to be "honestly pro-European," and the environmentalists, who promoted a federal European entity without giving up their ideological identity. The Socialist Party pleaded for a "federation of nation-states." The UDF, Les Verts, and the PS supported drafting a European constitution that would be, in essence, a mere sum total of previously implemented treaties and would serve to clarify the respective competences of the EU and the member states. Meanwhile, the Socialists had to decide on a common ticket in negotiation with their allied parties, the left-wing Radical Party (PRG) and the MDC led by Jean-Pierre Chevènement, whose European leanings were more restrictive and inspired by left-wing republicanism, which supported a "community of nation-states." During the negotiations, the MDC succeeded in convincing the PS to defend the so-called unanimity rule instead of the qualified majority rule for decision making in European institutions. The French president's position tended to support a less radical view of European integration. For example, on March 2, in his Parliamentary address made during the debate over ratification of the Amsterdam Treaty, the president spoke of a "United Europe made up of states" and tried to reconcile national and European ideals. During the campaign, the RPR-DL ticket regularly referred to Jacques Chirac's European policy. The PCF's position abandoned its overt hostility to Maastricht and aimed at preserving the current level of EU integration, unless efforts were to be made at developing European social policies. To preserve the national governments' privilege to veto the Council and to restrain the Commission's powers and reinforce the Parliament translate accurately the preferences for an intergovernmental logic and the democratization of European institutions. The RPF sovereignists headed by Senator Charles Pasqua and Philippe de Villiers patently favored a return to and greater control over national sovereignty as well as a step back from European integration in every sector; they made reference to a simple "association of states." The extreme right-wing tickets expressed their permanent anxiety over the "the national preference." Dreaming of a "new Europe," the Lutte Ouvrière and the Revolutionary Communist League (LCR) produced a common Trotskyite ticket. There were two kinds of electoral platforms. On the one hand were the broad orientations contained in the RPF, radical right, and radical left tickets. On the other hand, positions concerning common policies (agriculture, fishing), local politics (European regional projects), and European activities dealing with the environment, culture, and education were found in the tickets supported by the mainstream parties. We need here to draw a distinction between the platform of the environmentalists and the UDF, who felt the need to address every question possible and to be innovative in supporting pro-European positions, and the platform of the PS and RPR, which were shorter, more vague, and more moderate in their stance toward European construction.

In any case, there is no use denying that the campaign got buried in the news. It would be wrong to consider only the last two to three weeks of the campaign, even if this period was characterized by a rise in media coverage and a fall in voter indecisiveness. In reality, it had taken months for the electoral forces to find their balance, and the news was constantly updated and repeated to better orient the elector. According to Le Figaro of May 15 and Le Monde of May 18, the campaign began nine months before the ballot. Since the fall of 1998, there had been an overlapping of the national political life as shaped by the political parties and considered within the framework of the European elections, the course of European affairs, and the vagaries of international politics. To this effect, contributing factors included Daniel Cohn-Bendit's early candidacy; Pasqua's defiance toward the RPR; the UDF's reticence toward a common ticket with the RPR, and their final decision to stand alone; Bruno Mégret's takeover of the FN-MN; disagreements between Chevènement and Cohn-Bendit; negotiations held by the PS and its allies; ratification of the Amsterdam Treaty in January 1999; and the Santer Commission resignation in March. These events were all part of the campaign and constituted its initial phase. The campaign almost came to a full halt at the beginning of the NATO military intervention in Kosovo on March 24. Soon thereafter, the French agenda entered into a phase during which the media were saturated with other issues: the Kosovo refugees, the air force operations, and the diplomatic negotiations. When Philippe Séguin resigned from the RPR leadership, light was shed anew upon electoral news, yet the electoral platforms had great difficulty in making their campaigns seen and heard. This second phase was equally characterized by an administrative crisis: Bernard Bonnet, the *préfet* in Corsica, was charged with illegal use of law enforcement forces. As the opposition parties took advantage of the affair, it meant that public attention was engaged until the beginning of June. It was not until the first two weeks in June that pressure stemming from the Kosovo and Corsican affairs was eased, thereby giving way to the development of the final phase of the European campaign. The situation was further accommodated by the Köln European summit and the Blair-Schroeder joint declaration on a "third way" for Europe. The June 1 edition of *Le Monde* confirmed the difficulty that the campaign had confronted in attempting to overcome poor visibility amidst an overcrowded news environment: Journalists were searching for "substantiated proof of an ongoing electoral campaign" after several false starts caused by the NATO military intervention, Séguin's resignation as leader of the RPR, and the grandiose European Socialist meetings.

Campaign Perceptions and Electoral Mobilization

The mobilization for the European elections was a noncampaign for some, an antigovernment campaign for others. We believe the campaign was marked by a failing Europeanization even in the presence of certain signs that change occurred, such as the organization of international meetings and the elaboration of common platforms, the media's interest and the polls taken in other EU countries, and a greater priority granted to European issues in the French electoral debate. How did the electorate perceive the campaign? Relying on several polls conducted throughout Europe, we can compare the rate of interest for the campaign with the importance of national motivations and the most influential issues bearing on the ballot (see table 4.2).

The BVA poll shows the electorate in different countries with a similar ranking of issues. The pertinent issues are a mix of long-range concerns (employment rates) and immediate problems such as the Kosovo crisis. But is it really surprising to learn that voters were more interested in national questions than in European politics? Public opinion merely reflected (V. O. Key would say "echoed") the message of the political parties that tried to use the European elections to boost their national support. There is no question that a lack of interest was the predominant feature of the entire campaign. According to Sofres polls, during the four earlier European elections held in France, public interest in the elections had on average been at least 54 percent about three weeks before the vote. This rate was close to 38 percent in 1999.

Table 4.2 Campaign perceptions

BVA Survey (May 12–22)	Germany (%)	France (%)	Italy (%)
Interesting campaign	38	25	27
Uninteresting campaign	57	67	55
EU issues	43	31	22
Domestic issues	52	58	58
Employment	75	78	60
Kosovo	55	44	47
Taxes	50	33	35
Environment	33	30	18
Immigration	34	21	17

It is thus proven that in the four countries studied, the campaign was able to attract visibility during the last weeks. We can also shift the time period and then look at the polls' findings. The Ifop polling institution organized a survey the day after the elections to discern the reasons for the high French abstention rate. They received the following replies, in order of importance: "The campaign was not interesting" (52 percent), "There were no real European platforms" (28 percent), "The right-wing parties were divided" (25 percent), "European elections are not important enough" (19 percent), and no comment (5 percent). On election day, Sofres took a sample of nonvoters, who gave them another list of reasons that could explain their behavior. Forty-five percent responded that they "were not interested in these elections," 24 percent claimed that "no ticket was acceptable," 11 percent stated their desire to "express dissatisfaction with the Euro," and 7 percent made no comment. In a postelectoral poll designed for *Der Spiegel*, the following reasons were given for German abstention: a dissatisfaction with the policies of the government coalition (46 percent), a protest against European bureaucrats (37 percent), a sense of irrelevance of the European Parliamentary powers (34 percent), a lack of interest in the EU and its Parliament (28 percent), the weak importance of the elections (27 percent), and absolute ignorance of the elections (11 percent). In Germany, European elections had both a national and a protestor quality to them. The protest was aimed at the coalition government by support of the Christian Democrats and the PDS, the latter of which attracted voters from former East Germany who were dissatisfied with their lot after reunification. Because the electoral debate focused essentially on domestic issues, German and Dutch voters actually learned less about Europe during this campaign than they had 20 years earlier (Schoenbach, 1983).

All things considered, the reasons nonvoters gave led to a number of possible explanations for the low level of mobilization: the lack of campaign visibility, the responsibility of national political actors, and the negative evaluation of the EU. We have already discussed the extent to which the campaign was overshadowed by other "news" in the media. The public's reactions are thus an accurate picture of a hidden object. Our previous study of the 1997 French legislative campaign focused on the constraining cognitive conditions of electoral mobilization (Gerstlé, 1998). The 1999 campaign confirms that a certain degree of visibility is required to launch a mobilization process, as Blumler observed 20 years ago: The turnout rate increases in proportion to campaign activity (1983). The intensity of communication and support for the European Community constituted the two essential factors for explaining differences among national turnout levels. For instance, in Holland, where people widely supported the EC, there is no alternative but to attribute the low level of turnout to the weakness of the campaign. Similarly and in contrast to the British case, in Germany the strong support for the EC and a rather active campaign led to a high level of turnout.

Our second hypothesis, on the responsibility of national political actors, suggests that the electoral demobilization translated either a protest against the government or a disenchantment with political parties and their platforms, or even estrangement from politics. The coeditor of *Choosing Europe* (1996), Cees van der Eijk, strongly supports this argument when he attributes the record-high Dutch abstention level of 70.1 percent in June 1999 to a structural decline in interest in politics (*Le Monde*, June 15, 1999). A poll surveying the motivations of Dutch nonvoters reveals that 17.7 percent of them felt "no interest" in politics, 11.4 percent had "no interest in European elections," and 13.8 percent "did not have time." According to van der Eijk, the lack of interest appeared to be rooted in the sedative effect of the monotonous and unending message sent out by a coalition comprising Social Democrats and Republicans in a country where the market is increasingly dominant. The message read, "Citizens should not expect too much from politics." A third hypothesis concerns the negative evaluation of the EU and the growing distrust toward European institutions, particularly after the Commission crisis. Although the media's focus may have estranged many voters from European issues, notably in countries such as the Netherlands, where political corruption was a widely advertised theme, many surveys taken throughout the EU during the campaign showed that this deterioration of the European image did not instigate national retrenchment. On the contrary, pro-European feelings have been spreading,[1] and public opinion indicates a wish that greater European integration occur.[2] This tendency does not support the interpretation according to which weak electoral mobilization is a consequence of EU rejection. More convincing accounts for weak mobilization are the poor visibility of the campaign and the predominant national bent of campaigns, which were shaped by political parties and subsequently taken up by the media. If we were to limit our explanation to one exclusive factor, it would be the limited visibility of the campaign. In this respect, our argument is linked to the recent development of this factor in comparison with the preceding elections. It is evident that national issues had been as important, if not more, in former European campaigns and so cannot illuminate the reasons for the recent and dramatic drop in electoral mobilization.

The decline in electoral mobilization did not appear to be caused by an EU rejection (the EU has in fact stimulated expectations for integration). Nor can it be linked only to domestic instrumentalization of the European elections, often considered as "second-order elections." Given this context, the low level of voter turnout would be more accurately the outcome of a faltering Europeanization for this kind of campaign, due to the convergent effects of the international climate and an overbooked national agenda or one that was strategically active, such as in the German case, which overshadowed European affairs in the public's attention.

Notes

1. Ipsos survey (*Libération, La Stampa, Die Zeit, Le Soir, Publico, To Vima*) realized between May 7 and May 21 in 12 countries.
2. Louis Harris/*Le Monde* survey realized between may 6 and May 22 in eight countries.

References

Blumler, Jay G., ed., *Communicating to Voters*, 1983. London: Sage.

Gerstlé, Jacques, "La dynamique nationale d'une campagne européenne," pp. 203–227, in P. Perrineau, C. Ysmal, *Le vote des Douze. Les élections européennes de juin 1994*, 1995. Paris. Presses de Sciences Po.

Gerstlé, Jacques, "Dissolution, indifférence et rétrospection," pp. 55–76, in P. Perrineau, C. Ysmal, *Le Vote Surprise. Les élections législatives des 25 mai et 1er juin 1997*, 1998. Paris: Presses de Sciences Po.

Schoenbach, Klaus, "What and How Voters Learn," pp. 299–318, in J. G. Blumler, ed., Communicating to Voters. 1983, London, Sage.

Semetko, Holli, ed., "Symposium on Media and Politics in the Netherlands," *Political Communication*, 15, (2), 1998.

Semetko, Holli, and Klaus Schoenbach, "Parties, Leaders and Issues in the News," pp. 72–87 in S. Padgett, Th. Saalfeld, eds., *Bundestagswahl '98: End of an Era?* 1999. London: Frank Cass.

Van der Eijk, Cees, and Mark N. Franklin, eds., *Choosing Europe? The European Electorate and National Politics in the Face of Union*, 1996. Ann Arbor: University of Michigan Press.

Villa, Marina, "I temi e le strategie de la campagna pubblicitaria per le Europee 1989," in *Communicazione Politica*, no. 2, 2000.

PART TWO

A Europe of Political Forces

CHAPTER 5

The Communist and Extreme Left Galaxy

Marc Lazar[1]

A nalyzing the results of the Communist vote in the European elections is becoming increasingly complicated with each suffrage. In fact, the relatively homogenous bloc that western European Communist Parties formed for so long first cracked, then exploded. Since then, the Communist galaxy or that which emerged from Communism has itself been undergoing an accelerated process of differentiation. The strategies deployed by its different components have diverged, and the distance between them has grown steadily. Communist Parties in northern Europe, traditionally weak except in Finland, have clearly broken with Communism. In Sweden and Finland, they have transformed themselves into parties to the left of the left, while in Denmark and the Netherlands, they have either imploded or been integrated into coalitions with other small, radical groups. In both cases these Communist Parties have entered the post-Communist era. In other countries, that is to say Greece, Portugal, Spain, France, Italy, Germany, Belgium, and Luxembourg, Communist Parties intend to remain loyal to a portion of their original project, which they are nonetheless trying to redefine. The antagonism between post-Communist and Communist parties, however, cannot obscure other differences that weaken this already disparate whole in rather original ways.

For example, on the question of Europe, not only the post-Communist parties of northern Europe but also the Greek and Portuguese Communist Parties, which are resolutely hostile to the European Union, are pitted against the Italian, Spanish, German, and French Communist Parties, which condemn European integration but, believing it to be irreversible, militate in

Table 5.1 Results of the Communist Parties or former Communist Parties in the European elections

	1979		1984		1989		1994		1999	
	% vote	deputies	% vote	deputies	% vote	deputies	% vote	deputies	% vote	deputies
Austria[a]	–	–	–	–	–	–				
Germany[b]	0.4	0	1.3	0			4.7	0	5.8	6
Belgium	2.6	0	1.5	0	0.5[c]	0	0.9[d]	0	0.7	0[e]
Denmark[f]	20.9	4	20.8	4	18.9	4	10.3	2	6.5	1
Spain[g]	–	–	5.3	3	6	4	13.4	9	5.8	4
Finland	–	–	–	–	–	–	10.5	2[h]	9.1	1
France	20.5	19	11.2	10	7.9	7	6.8	7	6.8	6
Greece[i]										
Interior	5.3	1	3.4	1	14.2[j]	3	6.2	2	5.1	2
Exterior	12.8	3	11.6	3			6.2	2	8.6	3
Italy	30.8	24	33.3	27	27.6	22	6.1[k]	5	4.3[l]	4
									2	2
Luxembourg	5	0	4.1	0	4.7	0	1.6	0	2.9	0
Netherlands	1.7	0	5.6[m]	2	7	2				
Portugal[n]	–	–	11.5	3	14.6	4	11.2	3	10.3	2
Sweden[o]	–	–	–	–	–	–	12.9	3	15.8	3

[a] No Communist list.

[b] In 1979 and 1984, in the Federal Republic of Germany, it concerns the DKP, which in 1984 ran on a coalition list with the pacifists. Since 1994, it concerns the PDS, which emerged from the former East German Communist Party.

[c] In 1989 for the first time, the Belgian Communist Party did not present a list in the French-speaking electoral college. In the Flemish-speaking electoral college, the Communist Party was represented in a coalition list.

[d] The Belgian Communist Parties were part of coalition lists in both the French and Flemish-speaking electoral colleges.

[e] In the French community, the Communist Party obtained 1.1 percent of the vote. In the Flemish-speaking college, the PvdA, the Labor Party, obtained 0.56 percent of the vote, and the PAB obtained 0.2 percent of the vote in the German-speaking electoral college.

[f] The Communist Party in Denmark has always been all in favor of an anti-European coalition. The result of this coalition is the one indicated in Table 5.1.

[g] The first European elections in Spain took place in 1987. The Spanish Communist Party, itself divided into numerous organizations, is a member of the coalition Izquierda Unida (United Left).

[h] European elections of 1996.

[i] The first European elections in Greece took place in 1981. In 1968, a split divided the Communist Party into two parties, the Communist Party of the Exterior (orthodox) and the Communist Party of the Interior (reformist) which in 1987 transformed itself into the Party of the Hellenic Left, then in 1994 formed the Left and Progress Coalition. In order to simplify matters, we have retained the party's original name.

[j] In 1989, the Communist Party of the Exterior and the Party of the Hellenic Left presented themselves together as part of the Left and Progress Coalition.

[k] For 1994, it concerns the Rifondazione Comunista, the Communist Refoundation. The Italian Communist Party disappeared in 1991.

[l] Communist Refoundation Party score. The other score is that of the Party of Italian Communists (PDCI), resulting from the split in the PRC that took place in the autumn of 1998.

[m] From this date, the Dutch Communist Party ran in European elections as a member of a Green coalition. The Dutch Communist Party dissolved itself in 1991.

[n] The first European elections in Portugal took place in 1987. The Portuguese Communist Party ran within the framework of a coalition.

[o] The 1995 elections.

Table 5.2 Evolution of scores of Communist Parties or former
Communist Parties, 1994–1999

	1994 % vote	1999 % vote	Change
Germany	4.7	5.8	+1.1
Denmark	10.3	6.5	−3.8
Spain	13.4	5.8	−7.6
France	6.8	6.8	=
Finland	10.5[a]	9.1	−1.4
Greece[b]	6.2	8.6	+2.4
Italy	6.1	6.3[c]	+0.2
Luxembourg	1.6	2.9	+1.3
Portugal	11.2	10.3	−0.9
Sweden	12.9	15.8	+2.9

[a]In 1996.
[b]Only the result of the KKE, which officially claimed itself to be
Communist, is taken into account.
[c]The total score of the PRC and the PDCI.

favor of a different type of Europe. Environmentalism is another source of
disagreement among parties that have converted themselves (the post-
Communist Parties), those that utilize the issue (the German and Portuguese
Communists), and those that pay relatively little attention to it (the remaining Communist Parties). Finally, the European elections of June 1999
signaled the appearance of a new left-wing challenger to certain Communist
Parties—the revolutionary extreme left that emerged in certain countries,
especially France. This development may well have future consequences
within the Communist Parties themselves.

The breakup of the Communist camp limits its overall possibilities of
carrying any global weight and, on the contrary, requires a more refined
analysis that identifies mutations of the northern European post-Communist
left, Communist Parties, and the extreme left (see tables 5.1 and 5.2).[2]

Mutations of the Post-Communist Left in Northern Europe

In the Netherlands and Denmark, the minuscule Communist Parties have
dissolved into Radical-Green coalitions that relentlessly fight the creation of
a unified Europe. Nevertheless, the results are mixed. Holland's libertarian
left (Groenlinks) grew strongly (11.9 percent of the vote compared with 3.7
percent in 1994), while that in Denmark lost almost four points.[3] In both,
the former Communist militants occupy only marginal positions.

The results of the new parties that emerged from Communism in the extreme north of Europe and that seek to construct an alternative to Social Democracy are nuanced. The Finnish left-wing alliance, principally organized around the former Communist Party, has lost votes and a deputy. On the contrary, the Party of the Left (VP) in Sweden has had one success after another. The VP has progressed notably during the past five years, given that its electoral capital has more than doubled (see table 5.3). This election confirmed the sociological characteristics of its electorate, which is more feminine than masculine and younger rather than older. Twenty-two percent of those between the ages of eighteen and twenty-two voted for the party, which comes in ahead of the Social Democratic Party in these categories. The VP attracts many more unemployed (22 percent) and blue-collar workers (more than one-fourth of whom voted for it) than it attracts white-collar workers and farmers.

Numerous factors explain this dynamic. On the one hand, the VP presents itself as a resolute adversary of European construction. It has obstinately demanded, for example, a new referendum on the question of Sweden's membership in the European Union and has condemned the euro. On the other hand, it has developed a left-wing critical stance of the Social Democratic policies tying the country to the European Union and modernizing the welfare state. The party sees itself as the unremitting defender of established social benefits and, during the election campaign, made employment its major issue. Finally, it has broken with Communist ideology and has largely opened itself to postindustrial themes, notably demands from environmentalists and feminists. Its electoral lists presented the same number of men and women candidates, and the party has been directed since 1993 by Gudrun Schyman, a very popular woman with a high media profile.[4] The success of the Party of the Left raises the question of its future and its eventual exportability. Should the VP remain a protest party, indirectly influencing the governmental and strategic orientations of the SAP, or, as Gudrun Schyman (who claim to capture 20 percent of the vote) has suggested, should it become a party that takes part in government, forming clear-cut alliances with the Social Democrats? The debate is open within the party. Moreover, can the VP become an example to be imitated in order to build a party to the left of the reformist left? For the moment, only its Finnish counterpart has opted for an identical strategy. Elsewhere, the vigor of the traditional Communist culture precludes this type of metamorphosis.

Table 5.3   Results of the Swedish VP since 1994

1994 National elections	1995 European elections	1998 National elections	1999 European elections
6.2%	12.9%	12%	15.8%

Communist Parties

European elections generally do not favor Communist Parties, all the less so given that the matter divides them rather profoundly. It has been pointed out that a line now divides the Communist Parties that remain hostile to European integration, such as the Greek and Portuguese parties, from the Communist Parties that fight against the prevailing vision of Europe but declare themselves pro-European. The great change within the Communist camp comes from the fact that the French Communist Party, which was inclined toward the former camp, has recently rallied to the latter.

Communist Parties became involved in the electoral battle by trying to make everything grist for its mill. Probably because the theme of Europe is not particularly advantageous to them, but also because the theme has become part of their policy, Communist Parties attempted to present an attractive electoral offer by paying great attention to the makeup of their electoral lists. The objective was obviously to try and attract new voters. The war in Kosovo gave them the opportunity to reunite and transform pacifism into a resource they hoped would pay off with voters. All of the Communist Parties immediately denounced NATO's intervention and, at first, for quite a long period, remained silent about Serbian crimes. Communist Parties were deeply involved in street demonstrations. Pacifism was virulent in Italy, Germany, Spain, Portugal, and Greece, at times resorting to slogans for their shock value. The protests were marked by a total condemnation of NATO as well as an unusually aggressive tone toward Social Democracy. In Italy, demonstrators brandished effigies of the left-wing Democratic prime minister Massimo D'Alema, who was caricatured with Hitler's features. In France this pacifism was somewhat more temperate. In effect, the French Communist Party quickly realized that public opinion condemned Milosevic, felt strong compassion for the people of Kosovo, toward whom it expressed an intense solidarity, and approved of the NATO bombings. In addition, certain non-Communists on the Communist electoral lists approved the West's intervention, a matter that aggravated internal dissension and the uneasiness of many militants concerning the validity of the makeup of the electoral lists. On the whole, as we shall see, this opposition did not serve the Communist Parties very well, except in Greece and perhaps in Germany. Nevertheless, the war in Kosovo attested to the persistence of anti-imperialism and of the wholesale, indeed visceral, rejection of the United States within the Communist culture.

Two of these parties, in Belgium and Luxembourg, represent only vestiges whose results are of no interest other than to archaeologists of politics. In the French-speaking electoral college, the Communist Party's decision to present an electoral list for the first time since 1984 provoked a new and significant split within an anemic organization. In fact, some militants, particularly in

Brussels, preferred to take part in a Red-Green coalition, testimony to a recurring temptation within certain Communist Parties to favor this type of alliance. Under these conditions, the fact that the Communist Party obtained just over 1 percent of the vote is almost a miracle. In the Flemish-speaking electoral college, the Labor Party obtained almost 22,000 votes, or 0.56 percent. The Communists in Luxembourg, allied with the Trotskyites, improved their 1994 score slightly, nearing 3 percent of the vote.

The scores of the other Communist Parties, those that still count in the political life of their countries, obeyed three scenarios—catastrophe, defeat, and victory. Catastrophe for Izquierda Unida (United Left), a coalition of parties dominated by the Spanish Communist Party, which in 1994 obtained the best results of any European Communist Party. The Communist collapse (a loss of 1,284,417 voters, −7.6 points, and only four deputies instead of nine in 1994) is even more surprising in that it also occurred in the municipal and regional elections that took place the same day. The United Left focused its campaign on opposition to the war in Kosovo and, as such, missed its target. Only the list's official leader, the non-Communist Alonso Puerta, attempted to explain his party's position on Europe. The electoral rout provoked a crisis within the United Left's leadership. Julio Anguita tendered his resignation and was then reconfirmed in his post before having to retire definitely for health reasons.

The failure of Communist Parties has been appreciable, though less resounding, in Portugal, France, and Italy. On the occasion of the European elections, the Portuguese Communist Party tried to present a younger, less dogmatic image of itself by leading its Red-Green electoral list with Ida Figuiredo, a fifty-year-old economist and member of the party's central committee. Here also, the campaign heavily stressed its opposition to the war in Kosovo. But the PCP's score declined slightly in the election.

In France, the French Communist Party's secretary-general, Robert Hue, wanted to take advantage of the election to demonstrate the magnitude of the transformation he was seeking to bring about within his party. His list, "Move Europe," respected a double parity (men/women and Communists/non-Communists), the themes developed were more pro-European than in the past, and his budget was the largest of all the parties that took part in the elections, and four times higher than the Communist Party's budget in the 1994 race. But the result was inversely proportional to the physical effort, the creativity, and the financial investment. In the last several years, the PCF had been able to check the strong decline that started at the beginning of the 1980s, obtaining 8.6 percent of the vote in the presidential election and 9.9 percent in the 1997 legislative elections. But this time, with 6.8 percent of the vote, the PCF suffered a bloody defeat that probably reflected the deep turmoil of its voters. A large portion of the latter was highly skeptical of Communist participation in the government and of Europe (48 percent thought the European Union was heading the wrong way), while the makeup of the electoral lists and the positions taken by a number of non-Communists, especially regarding the war in Kosovo, confounded the most loyal militants and voters. The French Communist Party's score remained stable, even slightly lower, in comparison with its very poor showing five years earlier. Nevertheless, given the drop in voter participation, it gathered slightly fewer votes than in 1994, losing 150,000 voters. Finally, it now has only six deputies instead of seven. The French party continues to decline in its ancient industrial and rural bastions, while advancing slightly in lightly populated areas (western and eastern France). The sociology of its electorate remains unchanged, remaining weak among the young while gathering only 13 percent of blue-collar workers' votes, 8 percent of the vote of white-collar workers and midrange self-employed, and 6 percent of unemployed voters.[5]

In Italy, the results of the European elections give us the opportunity to analyze the balance of power between the two Communist Parties. In effect, the Communist Refoundation Party experienced a split in the fall of 1998. Today two parties antagonize each other with unusual vehemence. On one side is the Communist Refoundation Party (PRC), headed by former unionist Fausto Bertinotti. On the other is the Party of Italian Communists (PDCI), led by Armando Cossutta, leader of the former Italian Communist Party, who there represented the pro-Soviet wing but now militates for a reasonable, reasoned, and especially very political Communism. The European elections showed that the PRC carries nearly twice the weight of the PDCI (4.3 percent of the vote compared with 2 percent.) While the combined scores of the two parties is on the whole slightly higher than the Communist Refoundation score in 1994, it is nevertheless much lower than the 8.6 percent of the vote gathered by Communist Refoundation in the proportional part of the 1996 political elections. Furthermore, the division has left its mark and weakened the Communist camp in the Italian peninsula because neither of the parties received more than the symbolic 5 percent of the vote. According to a study by the Abacus firm, 56 percent of Communist voters in 1996 preferred this time to either abstain or cast blank ballots.[6] Lastly, the geographic structure of the vote for the PRC and the PDCI appeared more unbalanced than ever. The two parties together obtained 6.5 percent of the vote in the northwest, 5.2 percent in the northeast, 5.7 percent in the south, 4.3 percent on the islands, but 8.7 percent in central Italy, the traditional bastion of the Italian left.

The Greek and German Communist Parties, on the other hand, obtained indisputable victories, but with an important caveat. The victory in Greece seemed to have been unexpected and without much promise, while the

German victory was a step in keeping with a dynamic that has been noticed over several years, which could be confirmed in the future.

Indeed, the KKE benefited greatly from the war in Kosovo, which was massively condemned by Greek public opinion. The Communist Party denounced Europe and the PASOK government for its role in the war. It exploited hostility toward Americans and celebrated friendship with the Serbs. It even opened its electoral lists to non-Communist personalities known for their nationalist ideas, including a famous television journalist. It also drew up a highly negative assessment of the Socialists' economic and social policies. It was thus able to attain 8.65 percent of the vote (+2.4 points compared with 1994) and to strengthen its foothold among the working-class electorate. Furthermore, it clearly outdistanced the Progress and Left Coalition, its direct competitor that emerged from the Communist rift of 1968. Nevertheless, this victory turned out to be ephemeral. The Greek Communist Party obtained only 5.5 percent of the vote in the legislative elections in the spring of 2000. This contrasts with the 9 percent of the vote obtained by the Portuguese Communist Party, slightly better (a half point) than its score in the October 1999 legislative elections.

Along with the CDU-CSU, the PDS was one of the two winners of the European elections in Germany. It progressed 1.1 points compared with the 1994 elections and sent six representatives to the Parliament in Strasbourg. The PDS condemns the Europe of capitalism but claims to want to construct a different, more social and democratic Europe. Its position appears to be well received given that 52 percent of the voters think that participating in European integration is advantageous. In addition, its criticism of the Schroeder government's policies has paid off. But it was the war in Kosovo that allowed it to capture the pacifist space traditionally occupied by the Greens, because Green foreign affairs minister Joschka Fischer's support for military intervention wreaked confusion among his friends and electors. Nevertheless, despite the dynamism of PDS militants, the party remains fundamentally regional. Well established in eastern Germany, where it received more than 25 percent of the vote in Brandenburg, it remains almost nonexistent in western Germany outside the urban alternative circles of Bremen and Hamburg and despite an infinitesimal progress throughout the region. Its electorate presents the same tendencies—older, more rural than urban, with a high educational level—and has, especially, the strong impact of the unemployed within its ranks.[7] The divergence between the modernist and orthodox tendencies provoked a profound crisis within the PDS in April 2000.

Communist Parties are on the whole going through a difficult period. During the last elections, the principal parties (therefore excluding the Belgian and Luxembourg cases) oscillated between 4 and 10 percent of the vote, while five years earlier they had evolved between 6 and 13 percent.

Everything points to the worsening of their differences, be it on the question of Europe or on their conception of their relationship with Social Democratic and Socialist Parties and their projects. With regard to leftist reformist parties, Communist Parties adopt two attitudes. The first consists of a frontal combat in the old Communist tradition in which rivalry with Social Democracy frequently turns into hatred (the Greek, Portuguese, and Spanish Communist Parties and the Italian Communist Refoundation Party). It should nevertheless be noted that the electoral disappointments of the latter three parties have led them to adopt a more conciliatory position toward the PSP, the PDS, and the PSOE. In Spain, a common leftist program for the legislative elections was even signed between the Socialist Party (PSOE) and Izquierda Unida, an action that did not prevent the left from being roundly defeated and Izquierda Unida from suffering yet another rout in those elections of March 2000. The second form of behavior consists of accepting participation in coalition governments, following the example of the French Communist Party and the PDCI in Italy. Through this constructive attitude, the parties hope to strengthen themselves as the left-wing flank of the Socialist Parties. The German PDS would also like to carry out at the federal level a process similar to the one it has completed at the state level, where it already governs a *land* in a coalition with the SPD.

More generally speaking, three types of Communist Parties appear on the horizon: first, the orthodox parties that change only in appearance, such as the Greek and Portuguese Communists, even if the latter appear to have embarked on a prudent renovation "in the French style"; second, those parties that have chosen to bring about a change within a certain continuity, such as the German PDS, the French Communist Party, and the Italian Communist Party, three groups that are today very close but also very divided; finally, the parties that intend to bring together all of the Communist sensitivities by integrating them—for example, the Trotskyists. Examples of this third group include Luxembourg's Communist Party and Italy's Communist Refoundation Party, whichis becoming increasingly radical on all issues. This last party, having provoked the fall of the Prodi government, has ceaselessly attacked the government of Massimo D'Alema, whom it accuses of being at the service of capitalism and big business. It also did not hesitate to declare its agreement with the presumably theoretical portion of a text by the Red Brigades claiming credit for the assassination of an adviser to the labor minister in May 1999.

The Extreme Left

A notable fact is that after several years, the revolutionary extreme left—Trotskyist, Marxist-Leninist, and populist—has reappeared (see table 5.4). The European elections confirmed this trend.

Table 5.4 Extreme-left vote in 1999 (percentage of total vote)

Belgium	0.74%
France	5.2%
The Netherlands	5%
Portugal	3%[a]

[a]Includes the left bloc, formed by the Socialist Revolutionary Party (Trotskyist) and the Democratic Popular Union (Maoist), which together gathered 1.9 percent of the vote, and the PCPT/MRPP (Portuguese Workers Communist Party/ Reorganized Movement of the Proletariat), which obtained 0.9 percent of the vote, and the POLIS (Communist-Maoist) 0.2 percent of the vote.

In French-speaking Belgium, a list titled "Stand Up," (*Debout*) was ahead of the Communist Party by almost one point, with just under 2 percent of the vote. The list benefited from the dynamism of its leader, Roberto D'Orazio, a former steel unionist who became known during a strike in 1996 and 1997. An energetic, outspoken militant who believes in strong-arm tactics, he organized a demonstration of 70,000 people at the beginning of 1997 to defend the industrial site of Clabecq in Tubize (Wallonie). Excluded from the Socialist union FGTB, he founded the Movement for Union Renewal, which is largely dominated by the Belgian Workers Party (extreme left). His personality explains the fact that more than 46,000 Walloon voters chose to vote for his list.

In Portugal, a leftist bloc representing various extreme-left groups entered the elections. It included the PSR (the Revolutionary Socialist Party, Trotskyist) and the UDP (the Democratic Popular Union, of Maoist inspiration). This latter list obtained 1.9 percent of the vote, almost one point ahead of the PCPT/MRPP (the Communist Party of Portuguese Workers/ Reorganized Movement of the Proletariat Party). On the whole, this movement attests to the emergence of a fifth, structured force, a very minor one but one that will weigh on the future evolution of the Portuguese Communist Party. In the October 1999 legislative elections, the left-wing bloc gathered 2.4 percent of the vote and sent two deputies to parliament.

But it was in France that the extreme left achieved a tremendous advance. For the first time in their histories, the two principal Trotskyist organizations, the Communist Revolutionary League (LCR) and Workers Struggle (LO), elected five deputies, gathering more than 5 percent of the vote. Even more, the Trotskyists, with 910,946 votes, came in just behind the French Communist Party (1,192,155 votes). France therefore has the distinguished privilege, although in truth a bit outmoded, of witnessing the pursuit of the combat that Trotskyists and former Stalinist Communists have waged for more than 75 years. Workers Struggle and the Communist Revolutionary

League registered good results in formerly industrialized areas once dominated by the Communist Party (in northern France and in the Paris region, for example), and they came in ahead of the Communists in almost 15 major cities. Exit polls confirmed that a part of the Communist electorate preferred to vote for the Trotskyists. At the same time, Trotskyists advanced in rural regions and in those parts of the country where the left-wing vote was traditionally weak. All of this confirms a radicalized social and political protest vote. LO and LCR attracted 10 percent of blue-collar workers, 8 percent of white-collar workers and the midrange self-employed, and 6 percent of unemployed voters. The Trotskyist list, led by Alain Krivine and Arlette Laguillier, a militant who has acquired a certain notoriety after 25 years of public life, crystallized a vote that was unhappy with the left-wing coalition government and the construction of Europe.

In the Netherlands, the Socialist Party (SP) is a somewhat different case. Growing out of Maoism, it shed both Maoism and Stalinism in favor of a left-wing populism that blended materialistic and ecological themes. During the European electoral campaign, it continually denounced the European Union, emphasizing its gangrenous fraud and corruption as well as the record of the center-left coalition government.[8]

The June 1999 European elections show that there still exists a space, a small but real one, on the left flank of Socialist and Social Democratic Parties. The reformist policies of the governments have disappointed a portion of public opinion. The electorate to the left of the left is made up of working-class categories—workers, the unemployed, nonqualified public sector workers, and in general, with the exception of Sweden, older people. Nevertheless, the little bit of satisfaction that these voters represent for such parties should be strongly tempered. Unlike the Socialist and Social Democratic Parties, whose differences are real, as was illustrated by the episode of the Blair-Schroeder manifesto, but whose positions on Europe converge, the Communist and former Communist Parties are increasingly torn by centrifugal forces. The position of the United Left group in the European Parliament attests to this fact. While the group has increased in numbers—42 deputies in 1999 compared with 34 in 1994—it is characterized by its internal differences. First, five different political sensibilities are represented. These are the northern post-Communist left (6 deputies), the Communist left (24 deputies), the extreme left (5 French Trotskyist representatives and a deputy from the SP in the Netherlands), the Euroskeptical left (the two Greek deputies from the DKI), and the radical and modernist left (the other two Greek deputies from the former Greek Communist Party of the Interior). Furthermore and more importantly, this group, led by the French Communist Francis Wurtz, is profoundly divided over all of the issues concerning Europe: between the fierce enemies of Europe (the Nordic deputies, the Portuguese and Greek Communists and their allies in the DIKKI, the Trotskyists of the Communist Revolutionary League and Workers Struggle, the Dutch of the SP), about 19 deputies in all, and the 24 other elected deputies who are more pro-European (the Greeks of Synaspismos, and the German, Italian, Spanish, and French Communists). This is to say that during the entire legislature, the group will be obliged to seek compromises that will not help to clarify the content of its proposals. If, from the inside, the Communist movement is clearly identifiable, the Communist pole is pulled apart by powerful dissensions. All of this serves to confirm that while there are still parties that claim to be Communist, Communism itself appears increasingly elusive and constitutes a strong motive of discord.

Translated from the French by Eduardo Cué.

Notes

1. Professor of political history and sociology at the Institut d'Etudes Politiques of Paris, and dean of the graduate school.
2. This article benefited from the active collaboration of a certain number of colleagues. I would like to give warm thanks to Pascal Delwit (Brussels), Svante Ersson (Umea), Yves Léonard (Paris), Gerassimos Moschonas (Athens), Patrick Moreau (Munich), Jukka Pastella (Tampere), Luis Ramiro (Florence), and Gerrit Voerman (Groningue), who made facts available to me as well as their analyses on, respectively, Belgium, Sweden, Portugal, Greece, Germany, Finland, Spain, and the Netherlands. All my gratitude goes also to Mr. Philippe Poirier, scholar at the Gabriel Lippman Public Research Center in Luxembourg, for having provided me with comparative data on the June 1999 elections.
3. The result of this list is analyzed elsewhere in this book.
4. See Svante Ersson, "Le Parti de gauche et les élections suédoises de 1998," *Communisme*, no. 57–58, 1999, pp. 35–41.
5. See Marc Lazar, "La gauche communiste plurielle," *Revue française de science politique*, no. 4–5, August–October 1999, pp. 695–705.
6. *La Repubblica*, June 16, 1999.
7. For a recent study of the electoral characteristics of the PDS, see Patrick Moreau, "La situation électorale et politique du Parti du socialisme démocratique," *Communisme*, no. 57–58, 1999, pp. 7–34.
8. On this party, see Gerrit Voerman, "Das Gespensi des Kommunismus-Fine verblassende Erscheining. Gegenwart und vergangenheit des Kommunismus in den Niederlanden," in Patrick Moreau, Marc Lazar, Gerhard Hirscher (eds.), *Der Kommunismus in Westeuropa. Niedergang oder Mutation?* Landsberg: Olzog, 1998, particularly, pp. 515–518.

CHAPTER 6

The Disillusionment of European Socialists[1]

Gérard Grunberg and Gerassimos Moschonas

The PES on the Campaign Trail

The creation of the Party of European Socialists in November 1992 marked a new stage in the process of cooperation among Socialists of the European Union. In fact, given the stagnation of the previous phase, the PES contributed to reorganizing and heightening cooperation within European Socialism. The PES imposed itself and was gradually recognized as the unchallenged organizational center of coordination among Socialists at the European Union level, bringing a new dynamic to regional Social Democratic "integration." More homogenous than the Union of Socialist Parties of the European Community (UPSCE), founded in 1974, the PES is today more coherent and better equipped than its traditional partner and adversary, the EPP (European People's Party), to carry out "effective" actions within the European institutions. The PES's political influence has grown (especially through the party leaders' conferences) and its authority has been more clearly asserted and consolidated at the European level.[2] Yet, despite its increased strength, the PES, like all pan-European parties, is in reality only a "proto-party," a term indicative of a restricted partisan profile, even an elliptical one, and clearly incomplete.[3] The reinforced PES remains a weak integrative institution.[4]

Nevertheless, if the presence of the PES within the institutional elites of the European Union is increasingly felt and has acquired a certain "obviousness" or "visibility" (if only via the ambiguous route of Socialist summits), it is far from having crossed the threshold of visibility within European societies. The presence of the PES in the media (as well as that of all the pan-European parties) is marginal (although clearly more evident than in the past), and its influence on the political life of European countries practically nonexistent. The PES does not yet have the image and recognition of a real actor in European public opinion. It is therefore incapable of attracting the attention of the media and of mobilizing support ("identity-based" support and "systemic" support) on European issues.[5] Its visibility remains too weak even during campaigns for the European elections. The elections of 1999 easily confirmed this assertion.

The Manifesto

The PES's manifesto for the June 1999 elections was prepared—a "reassuring" enterprise or simply an ironic twist?—under the aegis of Robin Cook of Britain and Henri Nallet of France. Adopted by the Milan Congress on March 1, 1999, it was structured around four major themes: a Europe of jobs and growth, a Europe that puts citizens first, a strong Europe, and a democratic Union that works better. It also made "21 commitments" (*Manifesto of the PES*, 1999).

The text joins the long tradition of European party "federations" of all political shades that consists of producing documents "which are . . . bland, offering little more than platitudes . . . (and) little in the way of hard policy proposals."[6] The Socialist program, written in broad general terms that blur its European message, does not contain a single concrete promise. All of the questions that could raise objections (for instance, concrete measures to promote employment, the reform of the European budget and of European institutions, and the enlargement of the European Union) are either evaded or, more frequently, treated in extremely vague terms.

As a result of the consensual logic that governed its drafting, the document does not differ from other manifestos adopted in the past either by the PES or by the UPSCE. "The fact that the procedure of putting together such programs," writes Gilbert Germain, "obligates the representatives of the member parties, in the framework of the platform committees created by the federations, to confront their points of view, to better perceive each other's positions on certain precise themes, and to jointly seek a consensual base acceptable to everyone, even if it is a *minimum minimorum*, appears to us to be a determining factor in the evolution toward greater cohesion among political families on a European scale, as much on the conceptual level as in political practice."[7] Compared with the failure of the UPSCE to present a common manifesto for the 1979 elections (limiting itself to a simple "Appeal to Voters") the manifesto for the 1999 elections shows that an important consensual base currently exists in the PES. This base is constantly being expanded but it is fragile, and it will remain fragile as long as the procedure of putting a program in place is marked by the same profound contradiction. Electoral programs, as is justly underlined by Germain, are the ideal medium of debate among Socialists as well as the instrument of an evolution toward greater cohesion. At the same time, programs are also a perfect means of achieving a "hazy consensus" and, as a consequence, a narrowing of the debate. This inevitably produces a superficial cohesion. The constant production of program documents based on a *minimum minimorum* proves that there is still a long way to go before real cohesion is reached within European Socialism.

Furthermore, minimalist programs are not likely to be transformed into tools for action. Given the general nature of the promises made, it seems natural to us that the PES's last manifesto, just like those in the past, did not become a political, or even a purely verbal, instrument of a veritable European campaign. Devising programs with weak doctrinal and practical impact are not likely to truly engage member parties in an electoral action, a fact that was again demonstrated in June 1999.[8]

In the end, the national Socialist parties, as well as the PES, are faced with a double bind. They cannot become linked through an electoral program that makes firm commitments, and it is equally impossible for them to become linked through a program made up of excessively weak promises. It is also impossible for them to escape, by defining a middle line, this double impossibility (which is actually simply a problem of identity). Defining a middle road between an "extensive" program (which is the only one to create the basis for a consensus) and an "intensive" program (which makes a consensus difficult to achieve) is a procedure that has yet to be adequately developed, despite the growing ideological convergence within European Socialism.

"With regard to future electoral declarations," wrote Axel Hanisch, the former secretary-general of the PES, in 1995, "the PES could ask itself the following questions. To what point should we push the search for common declarations and therefore frequently the lowest common denominator? Would putting aside the viewpoints of certain isolated parties not enable us to express stronger political electoral objectives?" (Hanisch, *Rapport d'activités*, Brussels, 1995) These questions, asked after the 1994 elections, remain relevant following the 1999 elections. They also prove that putting into place "integrated" programs is a slow process, slower than was predicted by certain optimistic observers. There are two reasons for this. First, the absence of "integrated" programs corresponds directly to and reflects the weak internal "integration" of the PES. It is a game of mirrors. Second, this lack of integration is based on an institutional liberty. As long as the European party federations do not have to head a system of government, they can give priority to the unity of their organization (the "logic of membership") and, because of this, be imprecise, incoherent, and flexible regarding the formulation of their policies, and "permissive" concerning the promises made. Without institutional power, consensus is given priority. The consensus-based model of decision making has a weak "integrative" value and impact.

The Network Europe Group

The Network Europe group, created in 1998, answered the PES's need to replenish itself and experiment with new, more inventive forms of working collectively. Network Europe set itself the objective of infusing a new dynamic into the PES's internal and external communications as well as that of its group in the European Parliament. The clearly stated ambition of this communications network, which was probably inspired by the efficiency of the communications structures of the British Labour Party, was "to be the best."[9]

We, Network Europe, will be a dynamic and innovative team. . . . It is our task to add value to the information and communication activities within the European socialist family. Therefore we want to:

— Make the group more visible
— Make the members more visible
— Turn the discussions on European issues from a top-down and
  bureaucratic phenomenon to a relevant issue among the citizens of
  Europe . . .[10]

Network Europe proceeded to create three working groups, the Polling Network, the Media Network, and the Issues Network, with a specialist placed in charge of each of the three. These working groups multiplied contacts at the European level (including face-to-face meetings, with trips to most European capitals) with the press and with those responsible for opinion research and communication for the PES member parties.

A "Pan-European Poll" was carried out in December 1998 for Network Europe by Gallup International. The poll, which focused on the thematic priorities and public attitudes toward Europe in member countries, also tested the attractiveness (as well as the lack of attraction) of the different messages regarding European construction. The object of the research, which was not narrowly preelectoral (questions about partisan choices, voting intentions, and political parties were not asked), was to produce the information necessary to facilitate communication of the PES and the Socialist group in the European Parliament with the citizens of Europe. The results largely helped compose the guide that basically "structured" the informational work, both internal and external, of the Network Europe group. A number of conclusions emerge from the work carried out by Network Europe.

The group's and the party's Internet pages were clearly modernized and improved, as was the internal information system, through the development of the European Socialist Information Space (ESIS) on an intranet, which favored a better "structuring" and dissemination of information. Network

Europe organized and encouraged PES officials and experts to work together with member parties and Parliamentary national delegations. It also contributed to the establishment of the first European Socialist "group" in the areas of public opinion research, the media, and electoral management.

Nevertheless, this attempt at cooperation, which had not been seen on such a scale before the creation of Network Europe, remained limited to a small group of persons, an isolated microcollective that, after a flamboyant start, progressively lost its impetus and influence. In addition to the inherent difficulties in this type of collective work, it seems that the national priorities of the 1999 European elections became a "structural" obstacle that largely contributed to the group's loss of momentum and influence. Furthermore, the emphasis on and enhancement of Network Europe bypassed and depreciated the already existing communications structures in the party as well as in the Parliamentary group. This led to serious tensions, especially with those responsible for communication for the Parliamentary group, who even by Network Europe's own admission, were more familiar with the "corridors of the European Parliament." These tensions harmed Network Europe's overall communications effort as well as its range of influence. Finally, it is interesting to note that in reading the Manifesto, the political propositions of Network Europe were not really "incorporated" into the document detailing the PES's official program. This fact, an implicit disavowal of the work carried out by the group, is a bad omen for its future.

In short, the PES did not take the initiative of a truly European electoral campaign, a campaign that, beyond its national aspects, would have enhanced both the visibility of the party and the transnational themes it represented. The 1999 campaign was therefore not a truly transnational experience. It did not really put the PES on a more ambitious road and did not foster coordination of important joint actions despite the meritorious effort of the PES's leadership in multiplying contacts and encouraging its different components to work together. Certainly the party tested new, resolutely "modern" forms of communication and publicity. The impact of these innovations, nevertheless, was rather disappointing, and their effect, after an initial success, faded as time went on. Furthermore, the failure to adopt a common acronym and common colors (for example PES-PS, PES-PASOK), a consistently delayed decision,[11] was a sign of the weakness—even at the level of symbolic semantics—of the European dimension during the last election campaign. The party was not able to overcome the barrier of its "lack of visibility" and proved itself incapable of realizing the objective established at the Malmö Congress (1995), that is to say, "to pass from the role of internal coordination to external representation, promoting the public role of the PES."[12]

An Honorable Defeat

For the Socialists, the 1999 European elections were those of lost illusions. The myth of a Socialist Europe forfeited a great deal of its force. The Socialists, who govern 11 of the 15 countries of the European Union and participate in the government of a 12th member, saw the relative majority they held in the European Parliament escape them in favor of the right. Politically, the Socialists clearly lost these elections. From an electoral point of view, the defeat was more limited. When compared with the results of the previous European elections (table 6.1), a general downward trend does not appear. The Socialists advanced in five countries and were set back in ten. In five of the ten, the retreat was less than 2.5 percent. And it was greater than 5 percent in only one country, the United Kingdom, where the Socialists collapsed by losing 15 points.

Compared with the scores obtained in the legislative elections that preceded the 1999 European elections, the Socialists lost ground everywhere in 1999, but in greatly varying proportions (table 6.2). The 1994 European elections showed that the Socialist Parties in power lost votes everywhere compared with the preceding legislative elections, especially when they had been

Table 6.1 The Evolution of Socialist scores in the European elections of 1994 and 1999

	Socialist Scores		
	1994*	1999	Difference
Portugal	34.9	43.1	+8.2
France	14.5	22	+7.5
Spain	31.1	35.3	+4.2
Austria	29.2	31.7	+2.5
Denmark	15.8	16.5	+0.7
Luxembourg	24.8	23.2	−1.6
Italy	20.9	19.5	−1.4
Germany	32.2	30.7	−1.5
Sweden	28.1	26.1	−2
Ireland	11	8.8	−2.2
Netherlands	22.9	20.1	−2.8
Finland	21.5	17.8	−3.7
Belgium	22.3	18.6	−3.7
Greece	37.6	32.8	−4.8
United Kingdom	42.7	28	−14.7
Averages of the means	26	25	−1

*Or 1995–1996 for the three new EU members.

Table 6.2 Evolution of the Socialist vote since the next-to-last legislative elections
(percentage of ballots cast)

	Legislative elections preceding the 1994–1996 European elections	1994–1997 European elections	Legislative elections preceding the 1999 European elections	1999 European elections
Socialist parties in power at least prior to the next-to-last legislative elections				
Austria	34.9	29.2	38.1	31.7
Belgium	25.5	22.3	24.5	18.6
Denmark	37.4	15.8	36	16.5
Greece	47	37.6	41.5	32.8
Finland	28	21.5	22.9	17.8
Netherlands	24	22.9	29	20.1
Sweden	45.4	28.1	36.6	26.1
Socialist parties in power prior to the last legislative elections				
Germany	33.5	32.2	40.9	30.7
France	19	14.5	27.9	22
Italy*	22.6	20.9	21	19.5
Portugal	29.3	34.9	43.9	43.1
United Kingdom	34.4	42.7	43.2	28
Socialist parties in the opposition since last legislative elections				
Spain	38.7	31.1	37.5	35.3
Ireland	19.3	11	10.4	8.8

*PDS + Socialists.

in the opposition before those legislative elections, while they progressed or were only slightly set back when they were in the opposition.[13] We analyzed the 1999 elections in the same manner.

We distinguished three groups of countries. The first group is composed of the seven countries where the Socialists have been in power since at least the next-to-last legislative elections. The second group includes the five countries where the Socialists have been in power only since the last legislative elections. The third group includes the two countries where the Socialists have been in the opposition since the last legislative elections (table 6.2).[14]

In the first group of countries, the Socialist setback in 1999 is high compared with the legislative elections, between five and ten points. But compared with the losses observed between the European elections of 1994–1996 and the legislative elections that preceded these elections, the differences are slight, with the exception of the Dutch Socialist Party, which suffered a more decisive loss in 1999 than in 1994. In this group of countries, where the Socialists have been in power for many years, the European elections play the role of by-elections, providing some voters the occasion to sanction the party in power. Given the results, it is not possible to conclude that the Socialist Parties suffered a serious setback.

In the second group of countries, where the Socialists returned to power in the legislative elections that preceded the 1999 European elections, the setbacks were particularly significant, in keeping with tradition, in two countries, Germany and the United Kingdom (respectively ten and fifteen points), and relatively strong in France. On the other hand, the losses were less important in Portugal and Italy.

Finally, the elections confirmed that in the third group of countries, the parties who have recently gone back to the opposition suffered less significant losses in the European elections than when they were in power. This is the case in Spain and Ireland. The setback for the Socialist Parties in those countries between 1999 and the preceding legislative elections is around two percentage points.

Were the 1999 Socialist losses beneficial to other left-wing or environmental parties? Table 6.3 shows first that in 1999 the proportion of Socialist votes among all left-wing and environmental voters put together varied significantly from one country to another. In seven countries, the Socialists were clearly dominant, gathering more than 70 percent of the total vote. This was the case of the United Kingdom, Austria, Denmark, Germany, Portugal, Spain, and Italy. In the other countries the Socialist vote varied between 45 and 61 percent.

The Socialist victories and defeats, calculated on the basis of the number of ballots cast compared with that of the preceding legislative elections, should first of all be considered in relation to the total number of votes for the left and the environmentalists. Table 6.3 shows a strong relationship between the evolution of the Socialist vote and that of other left-wing and environmental parties. The majority of the activity takes place within the "left" in the broad sense of the word, that is to say, taking into account the environmentalist vote. Nevertheless, there are a number of exceptions to this rule. In Germany, the strong Socialist decline does not correspond to a rise of the non-Socialist left. The same thing is true in Denmark. In the United Kingdom, Socialist losses are to be found only partially in the gains made by the environmentalists. Finally, in Italy the slight decline of the PDS is accompanied by substantial losses on the part of Rifondazione Comunista (RC), losses due essentially to the scission that took place within the party. On the whole, however, the main obvious shifts take place within the left.

If the same calculations are made, but this time by comparing the 1999 European elections with the preceding European vote, the relationship

Table 6.3 Difference in the Socialist vote and the non-Socialist left and environmentalist vote between the 1999 European elections and the preceding legislative elections[15]

	Socialist parties' difference between 1999 European elections and preceding legislative elections	Non-Socialist left and environmentalists' difference between 1999 European elections and preceding legislative elections
United Kingdom	−15.2	+6.3
Sweden	−10.5	+8.7
Germany	−10.2	+0.4
Denmark	−9.5	−0.4
Netherlands	−8.9	+10
Greece	−7.7	+5.7
Austria	−6.4	+4.4
Belgium	−5.9	+7.4
France	−5.9	+6.7
Finland	−5.1	+4.3
Spain	−2.3	−4.8
Ireland	−1.6	+1.4
Italy	−1.5	−5
Portugal	−0.8	+1.7

previously observed is not as strong. The Socialist setbacks have a different meaning because they involve two elections of the same type, both of them by-elections. In the five countries where the Socialist loss is more than two percentage points, the losses are accompanied by gains on the part of the non-Socialist left and the ecologists. In the United Kingdom, nevertheless, the losses by New Labour are clearly higher than the gains by the environmentalists. In Belgium, Finland, and the Netherlands, gains by ecologists could mean that the environmentalist parties have permanently established themselves as serious rivals of the Socialist Parties.

In all of the countries where the Socialist Parties suffered minor losses or improved their scores, Austria excepted, the non-Socialist vote declines, at times very significantly, as was the case in Spain, the only country, along with Ireland, where these parties have recently been returned to the ranks of the opposition. In all of these countries, the results of the 1999 European elections do not therefore appear to translate into a declining trend for the left on the whole. In general, Socialism appears to be in a difficult situation only in a small number of countries and does not seem to be truly threatened by the non-Socialist left and the environmentalists, even if the latter have obtained very good scores in numerous countries.

Table 6.4 Difference in the Socialist vote and the non-Socialist left and environmentalist vote between the 1999 European elections and the preceding European elections

	Socialist parties' difference between 1999 European elections and preceding European elections	Non-Socialist left and environmentalists' difference between 1999 European elections and preceding European elections
United Kingdom	−14.7	+6.3
Greece	−4.8	+8.3
Finland	−3.7	+4.4
Belgium	−3.7	+4.3
Netherlands	−2.8	+7.3
Ireland	−2.2	−6.6
Sweden	−2	−4.9
Germany	−1.5	−2.6
Italy	−1.4	−3.2
Denmark	+0.7	−1.5
Austria	+2.5	+2.4
Spain	+4.2	−7.8
France	+7.5	−2.4
Portugal	+8.2	−0.9

This first analysis of the overall tendencies for the Socialists in the European elections demonstrates a certain diversity in the situations that incite us to observe more closely the different evolutions in the various countries. In order to carry out these observations, we utilized the first typology, allowing us to distinguish the countries where the Socialists have been in power since at least the next-to-last legislative elections, the countries where the Socialists returned to power during the last legislative elections before the 1999 European vote, and finally the countries where the Socialists have been in the opposition since the preceding legislative elections.

The Socialist Parties in Power since at least the Next-to-Last Legislative Elections

In the seven countries that make up this group—Austria, Belgium, Denmark, Greece, Finland, the Netherlands, and Sweden—the level of Socialist votes has always been lower in European elections than in legislative races. The most important differences concern Denmark, where, taking into account the political importance of the issue of Europe and the strong division among public opinion on this subject, European elections have been the occasion for anti-European groups and political groups to mobilize the votes of those opposed to the creation of a united Europe. We have here the clear distinction between a legislative electoral structure and a European electoral structure. The People's Socialist Party, the Danish anti-European left, has not benefit from the European elections, while the Greens' presence has been marginal. It is the right-wing, or the anti-European populists, who have been the beneficiaries of this type of vote. In Austria, Finland, Belgium, and the Netherlands it was the Greens who made the strongest advances in the 1999 European elections compared with the preceding legislative elections. In Sweden, it was the anti-European left, along with the ecologists, who made substantial progress. In Greece, where the ecological movement does not exist, the Communist Party profited most from the anti-European vote, as did the DKKI, which emerged from the PASOK as a new dissident party.

A comparison of the results of the 1999 European elections with those of the preceding European elections show, as we have already seen, that the Socialist vote declined in all of the countries that make up this group, with the exceptions of Austria and Denmark, countries where the Socialists were already at a very low level in 1994. In Sweden, the Socialist decline did not correspond to an increase on the part of the environmentalists, who in fact lost votes, but to a slight progress by the post-Communist left and especially to advances by the right. In Finland, the Socialist decline was due to gains by the Greens. In Greece the Socialist losses can be easily explained by the particularly unfavorable political context for the government of Costas Simitis. In effect, following the Ocalan affair, then the European and American intervention in Kosovo, which provoked a very negative reaction on the part of the Greek population and gave rise to anti-American sentiment, the Communist Party, which is both anti-European and anti-American, progressed. In these conditions, the 33 percent of the vote obtained by PASOK appears more like a positive result, and the government does not seem threatened.

The situation is different for the Belgian and Dutch Socialist Parties. In both countries, especially in Belgium, it was the government itself that was sanctioned and seriously challenged. In the Netherlands, the government led by the Socialist Wim Kok resigned in the weeks before the vote because of a conflict that opposed the Socialists' two partners, the left-wing party D66, which wanted to introduce the procedure of referendum into the Netherlands, and the Liberal Party, clearly opposed to such a measure. The three parties that made up the government coalition suffered a clear reversal in the elections, dropping from more than 60 percent of the vote in the last legislative elections to less than 50 percent in the June ballot. The vote took place in an uncertain climate, with the Greens as well as the Social Christians strongly advancing. Having governed with the latter, the Socialists changed their alliances following the previous legislative elections, which marked a clear setback for the Social Christians, who had led the government for many years. Despite the formation in July of a new government that is the exact duplicate of the previous one, the gains registered by the Greens and the alliance of Socialists and Liberals could in time modify the country's political balance.

In Belgium, the situation is more serious, with the strain of conflicts between different communities in the background. The scandal of dioxin in chickens that broke shortly before the elections (the European and legislative elections took place the same day), contributed to the government's rout. The strong advance by the environmentalists and the decline of the Social Christians, who led the government in an alliance with the Socialists, provoked a small political earthquake. Dehaene's government resigned. In its place, a Liberal-Socialist-ecologist government led by the Liberals was named. The situation therefore resembles that in the Netherlands, with a Socialist Party stuck between its Liberal ally and an environmental movement that has become a prime political force. In Belgium, the Socialists appear to be in a very difficult situation.

Countries where Socialists went from the Opposition to Power during the Last Legislative Elections

Five countries are involved here—Germany, France, Italy, Portugal, and the United Kingdom. The evolution of the Socialist vote in this group varies greatly from one country to the other, leading us to analyze the situation in each country individually.

Italy. The decline of the Socialist left is relatively weak compared with the 1996 legislative elections, retreating from 21.1 to 19.5 percent of the vote (table 6.4).

If one considers only the PDS, which since February 1998 has been rebaptized Democratici di Sinistra, the evolution of the vote is less favorable for this party. With 17.4 percent of the vote, the worst score ever achieved, and a setback of three percentage points since 1996 and almost two points compared with the last legislative elections, the PDS is the weakest of all the

Table 6.5 The evolution of voting in Italy since 1994

	1994 Legislative elections	*1994 European elections*	*1996 Legislative elections*	*1999 European elections*
PDS	20.4	19.1	20.6	17.4
Socialists	2.2	1.8	0.5	2.1
Communists	6	6.1	8.6	4.3
Greens	2.7	3.2	2.5	1.8
Total	31.3	30.2	32.2	25.6

political parties in power in Europe. In 1994 the PDS was in the opposition. In 1996 the Olive Tree coalition was led by a Christian Democrat, Romano Prodi, who was prime minister until the end of 1998, when the center-left government was put in the minority by the majority of the Communist Refoundation (RC) group and the right-wing opposition.

Despite this difficult situation for the left and the environmentalists as a whole, who with 25.6 percent of the vote reached their lowest score in Italy's electoral history since the end of the Second World War, assessment of the PDS's mediocre result should be mitigated. Numerous elements should be taken into account in order to interpret the PDS result politically. First, it was the first time that a government led by former Communists, a government that had been in place since the end of 1998, was confronted with the sanction of universal suffrage. Second, the PDS had to face the right and at the same time show that the center left remained in the majority. The latter gathered 41.2 percent of the vote against 38 percent for the Pole of Liberties. Despite the progress made by Berlusconi's party, the center left resisted well, and its political legitimacy was not challenged. Moreover, Massimo D'Alema feared the launch of Romano Prodi's new party, the Democrats, whose chief had not forgiven the ambiguous role played by the PDS in placing his government in the minority in 1998. The 7.7 percent of the vote gathered by the former head of government, while constituting a significant capital of votes, was for him a relative disappointment, and his attempt to again balance the center–left coalition was a partial failure. Finally, the RC, the PDS's enemy, suffered a notable setback as a result of the scission, dropping below the 5 percent vote margin. From this point of view as well, the PDS was reassured, whereas the RC has continued to progress since its creation. The PDS therefore clearly remains the principal left-wing force, even if its influence has been reduced and the PDS's electorate has shrunk. We should add that the war in Kosovo created a difficult climate for the left-wing government, with public opinion highly divided but globally hostile to a military intervention taking place off its coastline. Finally, the government came through this period without a major crisis despite the permanent threats of disintegration that weighed on its majority.

France. From an electoral standpoint, the 22 percent of the vote scored by the Socialists is not very high. The decline compared with the 1997 legislative elections is appreciable, almost six percentage points. Any comparison with the 1994 European elections must be made with caution. Certainly, the almost seven point increase is the second largest in Europe after that of the Portuguese Socialist Party. But this increase must be seen in a relative light. In 1994, the Socialist list was challenged by a list composed of Jean-Pierre

Chevénement's allies, and especially by a list of radicals directed by the populist Bernard Tapie, who in reality was supported by President François Mitterrand. This list obtained 12 percent of the vote. Even if the comparison is open to discussion, the three organizations that ran for election separately in 1994 and that obtained a total of almost 29 percent of the vote were united under the Socialist banner in 1999 in a list headed by François Hollande. The Socialist gains in 1999 should therefore be analyzed in light of this important modification of the electoral offer. Nevertheless, the 22 percent of the vote garnered by the Socialists in 1999 was read politically as a Socialist success for numerous reasons, among them, firstly, the dispersal of the right into three different lists and the RPR's calamitous score of 12 percent of the vote; and secondly, the stagnation of the Communists at their low level of 1994 (6.8 percent of the vote), followed hot on the heels by the Trotskyist list (5.2 percent of the vote). Despite the presence of three lists of the plural majority, it gave the impression of unity, in contrast to the disunity portrayed by the opposition. Lionel Jospin's government in fact came out reinforced by the elections, especially given that the British and German Socialists registered severe defeats. In the conflict with Blair and Schroeder, which we will discuss further on, and their manifesto published on the eve of the elections proposing a third, social-liberal road in contrast to that of the French Socialist Party, the latter saw in its relative success the opportunity to block the ideological offensive of the two other major leaders of European Socialism. This success allowed Jospin to begin the second part of the Socialist legislature under favorable conditions.

The only dark spot for the French Socialists was the success of the Greens list, headed by Daniel Cohn-Bendit, the former May 1968 leader, which achieved an excellent score of almost 10 percent of the vote, compared with 2.5 percent in 1994. The Greens' score (the Socialist Party gathered just 50 percent of the total vote for the left plus the Greens in 1999, compared with 65 percent of the vote in 1997 and 85 percent in 1994) allowed them to increase their demands for greater representation within the government.

Nevertheless, Jospin, buoyed by his extraordinary popularity, should not fear major difficulties for the moment. The question is whether the Green Party, which has made significant electoral gains in the past, will be capable of confirming its success during the upcoming elections and settling in, as it intends to, as the second political family of the plural left. On the whole, the Socialists emerged from the European elections in a stronger position.

Portugal. The big surprise of the European elections was the extraordinary score attained by the Portuguese Socialist Party, even though it had been in power since the previous legislative elections in 1995. With 43.1 percent of the vote, or barely one percentage point less than its score in the 1995 legislative elections, the party, led by António Guterres, confirmed that it had become the leading Portuguese party. It progressed by almost ten percentage points since the previous European elections. How can such a success be explained? First, it should be recalled that the advance of the PSP had been interrupted since the end of the 1980s. With the arrival in 1992 of Guterres at the head of the PSP, the party profited from the strong trend of bipolarization of political forces in Portugal as well as from the inability of center parties to interpose themselves between the two major political forces, the PSP and the PSDP. Furthermore, following the retirement of Cavaco Silva, the PSDP's leader, the major center-right party that won crushing victories at the beginning of the 1990's has not found a leader of a similar caliber to replace him. In addition, the Socialist leader, helped by the economic development of the country, has recentered his party. A devout Catholic, he did not wish for an alliance on the left and preferred to occupy the center left, not fearing to appear as a "Portuguese Tony Blair."

Finally, Mario Soares, who led the Socialist list and was the PES's candidate for the presidency of the European Parliament in case of a Socialist victory, made it possible for the list he conducted to benefit from his immense personal popularity. The 1997 municipal elections, the first after the formation of the new Socialist government, had already proved a real success against all odds for a party in power. With 41 percent of the vote, the Socialist Party gained more than two percentage points and further increased its distance from the Social Democratic Party (the right-wing party).[16] The same phenomenon occurred in June, with the Socialists failing to be sanctioned by the electorate. The Communist left allied with the ecologists progressed only slightly compared with the legislative elections, increasing 1.7 percent but falling by one percentage point compared with the last European elections. The PSP has therefore imposed itself since its return to power as the dominant party on the left, and the European elections confirmed its advantage over the right. This case is a clear exception to the model used. The long series of electoral success at all levels reached its apogee with the Portuguese Socialists' clear victory in the legislative elections of October 1999, when they received 44 percent of the vote, the same score as in the previous legislative elections and within a hair's breadth of an absolute majority in parliament. In the midst of these victories, Guterres was elected president of the Socialist International in November of the same year, succeeding Pierre Mauroy. It should nevertheless be noted that the Communist extreme left and Greens campaign against the European Union and notably against the intervention in Kosovo allowed it to score a certain number of points in the face of a pro-European Socialism that adopted a cautious but nevertheless favorable position to the intervention.

Germany and the United Kingdom. There are several reasons for dealing with these two countries at the same time. It was in Germany and the United Kingdom that the Socialists suffered their most stinging defeats. Both governments, in an alternating system of two major parties, had been led by Socialist governments for short periods—two years for the Labour Party and only one year for the SPD. The two leaders, Tony Blair and Gerhard Schroeder, had chosen to publish a joint manifesto a few days before the election. In both cases, moreover, their defeat essentially benefited their rightwing rivals despite an honorable score by the Greens in the United Kingdom. Although in England, given the abstention rate, the Labour setback must be put into perspective, in Germany the strong comeback of the Christian Democrats and the chancellor's low popularity rating placed the Red–Green coalition government on the defensive until the scandal that struck the CDU at the start of the year 2000.

The slight loss suffered by the SPD compared with the preceding European elections can be explained by the fact that in 1994 the SPD, in the opposition, deeply divided, and without real leadership, registered a particularly poor score. Furthermore, the Greens were in the government for the first time in 1999 and they therefore also paid the price of the government's unpopularity. The results of the European elections by *land* show a disturbing evolution for the Social Democrats. In effect, while they advanced in Bremen and Hamburg, they lost badly in Brandenburg to the Christian-Democrats, while in Bavaria the CSU gained formidable hegemony with an absolute majority. The governmental coalition was defeated by the CDU-CSU opposition in 11 of the 16 German *länder*.

In the United Kingdom the disaster was even greater in that it was unexpected. The polls predicted a good score for Labour, predictions that were foiled by a high abstention rate that reached 80 percent. Labour voters remained at home while the anti-Europeans were strongly mobilized. Labour's collapse may not have major electoral consequences in the next legislative elections despite the Conservatives' unexpected score, but this electoral setback was a political defeat for Tony Blair in several important respects.

The most consequential was that of the euro. Certainly Tony Blair understands that he made a mistake in painstakingly avoiding discussion about Europe and especially about the Euro in these elections, therefore failing to mobilize his camp, and by thinking that his Churchillian or Thatcherist attitude on the war in Kosovo and his strong popularity would be enough to win the elections. But even if he implicates himself more during the promised referendum on the Euro, it is clear that mobilizing Labour voters on this theme will be more difficult than expected. The referendum on the Euro appears likely to be an even more formidable test than originally thought. All of this is very detrimental to Tony Blair, who intends to play a leading role in European politics.

The election results also show that workers belonging to the Labour Party registered the highest abstention rate in the European elections. From the point of view of electoral geography, the Labour losses, compared with previous European elections, were greater in the old industrial north and in Wales. This can certainly mean that the working class is not very interested in European issues, but it can also mean that the Socialists' popular base is disappointed by two years of Labour government.

The defeats of the British and German Socialists also had consequences for the ideological debate within European Socialism. In effect, the European elections took place at the same time that Tony Blair was developing his offensive to make the Third Way the political and ideological platform of Social Democracy. The SPD's victory in Germany allowed him to reinforce his links with Schroeder, in whom he had found an ally. Despite the fact that the European Union's Socialist Parties had adopted a common manifesto for the June European elections during the PES Congress in Milan in May, Tony Blair and Gerhard Schroeder signed and published a manifesto entitled "Europe: The Third Way/Die Neue Mitte" just a few days before the vote. The manifesto, with a clearly liberal tone on economic issues, was in sharp contrast with the text adopted by the PES in Milan. Numerous Socialist Parties, especially the French, did not appreciate the Blair-Schroeder initiative, in either form or content.

The electoral defeats of the two leaders in the European elections had notable consequences for the evolution of the internal debate. The defeats weakened the two leaders' positions within European Socialism and favored a counteroffensive by French Socialists. The setback could now be used to show that the Third Way was not much of an electoral asset, whatever the real reasons for the defeat may have been, while the French Socialist Party had on the contrary registered a satisfactory electoral result. The Blair-Schroeder manifesto therefore appeared as the symbol both of Blair's offensive within European Socialism and of the British and German electoral defeat. The Socialist International Congress held in Paris in November 1999 allowed the anti-Blair forces to impose a synthesis entitled "The Declaration of Paris," which, while taking into account certain themes of the Third Way, was essentially a reminder of Social Democracy's traditional position, namely its critical attitude toward capitalism and its belief in the primacy of politics. Schroeder's own political difficulties in his country and within his party did not allow him to play a leading role at the Paris Congress. The Socialists' desire to maintain their unity did not transform the disagreements into confrontation, but the debate is far from closed.

Socialist Parties in the Opposition

Despite the slight losses it registered compared with the previous legislative elections that sent the PSOE into the opposition after 13 years in power in Spain, the Socialist score in the European elections was much higher than that expected by political observers—35.3 percent of the vote, just four and a half percentage points less than the Popular Party. The prediction that the Socialists would be defeated was based on two elements. The first was the crisis in the PSOE's leadership following the retirement first of Felipe Gonzalez and later, just a few weeks before the vote, of José Borell, the party's candidate to head the Spanish government in the event of a Socialist victory in the next legislative elections and the man who was to conduct the Socialist list in the European elections. The second element was the strong popularity of Prime Minister Aznar. Under these circumstances the relative Socialist success can be explained in two ways. First, the absence of a real Socialist leader paradoxically helped the PSOE in that the Popular Party did not know which Socialist personality to attack, while Felipe Gonzalez, who remained very popular, was nevertheless present in the Socialist panorama. Second, conforming to our model, the fact that the Socialist Party was in the opposition allowed it to reconquer a left-wing protest vote that in the previous elections had gone to the extreme-left list principally made up by the former Spanish Communist Party. In effect, the Unified Left (Izquierda Unida) list collapsed. It received 13.6 percent of the vote in 1994, 10.6 percent in 1995, and 5.8 percent in 1999, scores that led to the resignation of its leader, Anguita, a resignation rejected by the party. The PSOE's hegemony over the left is now stronger than ever. This position allowed it to propose a governmental alliance with Izquierda Unida for the legislative elections of the year 2000.

The autonomous elections, which took place at the same time as the European vote, were also positive for the PSOE, especially in Andalusia, Asturias, Catalonia, and Extremadura. The Socialists also benefited from the municipal elections, obtaining 34.2 percent of the vote in 1999, about the same as the Popular Party. The October 17 elections in the autonomous region of Barcelona confirmed the electoral good health of the Spanish Socialists, as they came close to winning.

The PES within the European Parliament

Following the 1999 European elections, the PES lost 34 seats compared with the 1994 elections, going from 214 to 180. This significant setback made them lose their relative majority in the European Parliament. The right's victory translated into an even larger right–left cleavage during the election of the Parliament's new president, because Mario Soares, the Socialist group's candidate, obtained only 200 votes against 306 for the right-wing candidate, Nicole Fontaine of France. The environmentalist candidate obtained 49 votes. The Socialists were therefore able to count on only their own votes and about 20 more from the remainder of the left. The loss of 34 seats was essentially due to the setback of British Labour. They in effect lost 31 seats, partly as a result of the reform that changed the voting from a majority to a proportional system. Actually, the Socialists lost 46 seats and won 12. The most notable increases concerned France (+6) and Spain (+3). The only significant setback, in addition to that of New Labour, was that of the German Social Democratic Party (−7) (table 6.6).

Table 6.7 shows the evolution of the respective portions of the Socialist national contingents within the Socialist group as a whole in the European Parliament. British domination collapsed and Pauline Green was replaced by the Spanish Socialist Enrique Baron as head of the group. The number of Spanish and especially French deputies in the group grew considerably. Today the German contingent is the largest, with 18 deputies, followed by the British contingent, which declined from 28 percent of the deputies in the Socialist group to 17 percent. The British are followed by the Spanish and French groups. The British and Germans, who together represented 47 percent of the Socialist group, today represent just 35 percent.

Table 6.6 Socialist seats in the European Parliament

	1994–96 European elections	1999 European elections	99/94 Difference
Germany	40	33	−7
Austria	6	7	+1
Belgium	6	5	−1
Denmark	4	3	−1
Spain	21	24	+3
Finland	4	3	−1
France	16	22	+6
Greece	10	9	−1
Ireland	1	1	0
Italy	19	17	−2
Luxembourg	2	2	0
Netherlands	7	6	−1
Portugal	10	12	+2
United Kingdom	61	30	−31
Sweden	7	6	−1
Ensemble EU	214	180	−34

Table 6.7 Percentage of Socialist seats from each country in the PES

	1994–96 European elections	1999 European elections	99/94 Difference
Germany	19	18	−1
Austria	3	4	+1
Belgium	3	3	0
Denmark	2	2	0
Spain	10	13	+3
Finland	2	2	0
France	7	12	+5
Greece	5	5	0
Ireland	0	1	+1
Italy	9	9	0
Luxembourg	1	1	0
Netherlands	3	3	0
Portugal	5	7	+2
United Kingdom	28	17	−11
Sweden	3	3	0
Ensemble EU	100	100	

Conclusions

1. Comparison of the results of the 1999 European elections with those of the previous European vote underlines the reductionist nature of the hypothesis diagnosing the overall electoral reverse of the Socialist and Social Democratic Parties. Certainly the Socialists, *taken on the whole as a political family*, were set back at the polls (although moderately), but the image of an overall decrease hides highly different, even opposed, tendencies at the national level. The analysis of the data highlights the diversity of the national cases (the Socialists fell back in ten countries and progressed in five compared with the previous European elections).

2. Nevertheless, the decline is across the board compared with the scores obtained during the legislative elections preceding the 1999 European elections, regardless of whether the Socialist Party was in the government or the opposition. Certainly being in power or in the opposition during the legislature (or legislatures) preceding the European electoral consultation is an important key to explaining the Socialist electoral performances. In the sense that the Socialists are in power almost everywhere, the phenomenon of an electoral decline is normal. But these elections only *partly* confirmed the theory that participation in the government constitutes an important factor in accentuating electoral losses (opposition status, according to this thesis, is a factor that slows the decline, or even fosters electoral gains), as is indicated by the satisfactory scores obtained by the Portuguese, French, Danish, and Austrian

Socialists. The *general* tendency to decline (compared with the performances realized in the legislative elections) shows that the aspect of a "by-election" weighs heavily, next to the factor of "participating in the government," on the electoral performance of the Socialists.

3. The last European elections showed once again that contemporary Social Democracy is an unstable, vulnerable electoral force. Such an electorally unstable force can be subjected to important and rapid losses and can, at times just as rapidly, reconquer lost territory. It is therefore incapable of being effective in a relatively stable manner for a long period of time. The Socialists of today, whether in their "leftist" or "rightist" version, whether in the southern or northern version, are highly sensitive and exposed to economic factors, to the convulsions of politics, and to the permanent reversal of situations.[17]

4. Through their initiative of launching a common manifesto on the eve of the European elections, the two British and German Socialist leaders not only involuntarily underlined the nonbinding character of the PES's documents outlining a program, they also took the risk of making the results of the European elections in their respective countries a stake in the internal debate of European Socialists. In effect, the electoral defeats of the two leaders in the European elections *contextually* weakened their positions within the European Socialist movement and favored a counteroffensive by the French, determined—at least theoretically—"to create an obstacle" to the Third Way. The European elections, through the occasion they provided the partisans of the Third Way to declare their positions, and at the same time by the effect of New Labour's and the SPD's electoral defeats, finally played a certain role in the crystallization of the internal debate and in the manner in which it developed and temporarily concluded. This role was nevertheless limited, not only because the "transnational" impact of a European election was by definition limited, but also because the convergences in European Socialism won largely over the divergences.

5. An overall view of the 1999 elections shows that the European identity proposed by the Social Democrats is still too weak to truly contribute to the formation of a "European conscience" in the sense of Article 138A of the Maastricht Treaty. At the same time, the weakness of the "European conscience"—considered not as a substitute for the national conscience, but, to employ the term of Dimitris Tsatsos, as a "second level" of politics and politicization directly linked to the existence of a *European commons*[18]—makes it difficult to propose an identity that does not use national vocabularies and does not orient itself toward a national clientele. The nation remains the center of partisan identifications, something that handicaps the transnational groups of political parties. This was confirmed once more by the last European campaign.

6. *Union,* wrote Guillaume Devin in 1989, remains fundamentally "the instrument of Socialist national policies, less in order to transcend them than to legitimize them through a common formulation."[19] Today the PES, the inheritor of *Union,* without being "the instrument" of national politics, remains a structure that has yet to find its place and its role in the continual and constantly topical shuttling between the "national" and the "supranational" level. Certainly this shuttling "only requires a balance to be found" and not a clear-cut division between pro-integrationists and Euroskeptics. The current PES is without doubt consolidated and reinforced as an organization that structures Socialist cooperation in the European Union. But it is still seeking to find a path for itself in a difficult space where the issue of frontiers dominates—frontiers between the national and the supranational and between the nations that make Europe and a Europe that is greater than the nations that constitute it, frontiers also between the "nationalistic" dimension and the international dimension of the most intimate tradition of European Socialism. In this space, where conflicting interests cross, from fragile loyalties to heterodox logic, the PES, like all pan-European parties, seeks its bearings. It is not surprising therefore that it is feeling its way, that it hesitates, that it advances slowly and frequently masked. The "frontier" is internal, it runs across the party, its leaders, and the party members that make it up. It traverses European Socialism in its entirety. This frontier nevertheless does not constitute a "front line" (as the etymology suggests). It is moving, fluid, indecisive, inconsistent, intangible, but real. In the end, European Socialism, wrought by contradictory tendencies and aspirations, finds itself with a strategy in limbo because it is caught between two identities. The 1999 elections only confirmed this intermediary state—of spirit and of strategy. The elections showed that a "Common Market of political parties"[20] has not yet emerged in Europe. In this sense, these elections, which confirmed what we already knew, were only a routine event. An interesting routine event.

Translated from the French by Eduardo Cué.

Notes

1. We would like to thank our colleagues who graciously provided us with information and shared their thoughts with us in order to write this chapter: Yves Léonard for Portugal, Luis Ramiro for Spain, Marc Lazar for Italy, John Crowley for Great Britain, Swante Ersson for Sweden, Gerrit Voerman for the Netherlands, and Pascal Delwit for Belgium.
2. Hix, S., "Political Parties in the European Union: A Comparative Politics Approach to the Organizational Development of the European Party Federations," *Working Paper,* Jan. 1995, Manchester, passim.

3. Soldados, P., *Le système institutionnel et politique des communautés européennes dans un monde en mutation*, Brussels: Bruylant, 1989, p. 231.
4. Ibid, 186.
5. On L. N. Lindberg's and S. A. Scheingold's distinctions between *identity-based* and *systemic* support, see Soldatos, 241.
6. Smith, J., *Europe's Elected Parliament*, Sheffield: Sheffield Academic Press, 1999, pp. 93, 96.
7. Germain, G., *Approche socio-politique des profils et réseaux relationnels des socialistes, libéraux et democrates-chrétiens allemands et français du Parlement Européen*, doctoral dissertation, Paris Institute of Political Studies, 1995, pp. 309–310.
8. The publication, just a few days before the June elections, of the Blair-Schroeder manifesto showed in spectacular fashion the "nonbinding" and "noncommittal" nature of documents outlining the PES's program. This manifesto, a text outlining the differences rather than the similarities within European Socialism, was badly received and caused friction (and the irritation of the French) in the PES (see infra).
9. "Quality and excellence," according to the terms of Network Europe, was the principle on which the group wanted to found its action: "To have the best management systems (internal communication networks, Intranet, information management and outreach strategy), the best European political web site, the best political story adaptable to national circumstances, the best European polling program, the best media management" (Network Europe, *Final Report*, internal document, Brussels, June 30, 1999, pp. 2–3).
10. Network Europe, *Final* Report, ibid., pp. 2, 6.
11. T. Beymer, the PES's secretary-general, appears convinced that the Socialists will be able to present a common acronym for the next European elections (interview with G. Moschonas, Brussels, November 1999).
12. *Activities Report of the PES, From Malmö to Milan*, PES, Brussels, 1999.
13. See Gérard Grunberg, "Le socialisme en difficulté," in Pascal Perrineau and Colette Ysmal, *Le vote des Douze, Les élections européennes de juin 1994*, Chroniques électorales, Presses de Science Po, 1995, pp. 39–73.
14. Luxembourg was excluded from this analysis because both the legislative and European elections took place at the same time.
15. The differences are calculated based on the percentage of votes cast.
16. See Yves Léonard, "Portugal: en attendant l'Europe," in *Les pays d'Europe occidentale* (ed. A. Grosser), 1998, Les etudes de la Documentation française, pp. 209–230.
17. See G. Moschonas, *In the Name of Social Democracy, The Great Transformation*, London: Verso, 2001 (forthcoming).
18. See Tsatsos D., *Des parties politiques européens? Premières réflexions sur l'interprétation de l'article 138A du traité de Maastricht sur les partis*, Brussels, p. 5.
19. Devin, G. "L'Union des Partis Socialistes de la Communauté Européenne. Le Socialisme en quête d'identite," in *Socialismo Storia*, Milan: Franco Angeli, 1989, p. 282.
20. Henk Vredeling's expression cited by Simon Hix (*op. cit.*, p. 2).

CHAPTER 7

The Greens in the 1999 European Elections: The Success Story

Ferdinand Müller-Rommel

The Greens inside the European Parliament: A Historical Review

Green Parties and alternative lists participated in all five direct elections to the European Parliament (EP) (Bomberg 1998: 102–125). In 1979, various ecological and small leftist movements nominated their own candidates, who received encouraging electoral support in the first European direct elections but no seat in the EP: 3.2 percent for the Grünen in West Germany, 2.3 percent for Agalev in Belgium, 1 percent for the Alternative Left in Luxembourg.

Prior to the second direct elections in 1984, a committee was formed in order to create a common electoral platform for the various Green Parties and alternative lists that existed in Europe at that time (Müller-Rommel 1985: 394). However, from the very beginning, a conflict emerged between the "purist" Greens, who adopted predominantly ecological concerns in their party program, and the "leftist" Greens, who supported an ecological strategy that also called for a radical change of "capitalist" society. Due to the ideological disagreement between the "leftist" Greens (from Germany, Luxembourg, and the Netherlands) and the "purist" Greens (from France, England, Ireland, Belgium, Sweden, and Austria), no common campaigning strategy was developed. In spite of this situation, several Green Parties and lists were surprisingly successful in the 1984 European elections: The Belgian and the Dutch Greens gained two seats each, while the German Greens received a total of seven seats in the EP.

Although the parliamentary rule allowed at least ten members of the Parliament from a minimum of three different countries to form a party group within the EP, the German and the Dutch "leftist" Greens (with nine seats) rejected establishing a "Green Party federation" with the "purist" Belgian Greens. Instead, the German and Dutch Greens looked for coalition partners among other left-wing and regional parties within the EP. After several rounds of negotiations, the "Rainbow Group" was formed as a Parliamentary group (Buck 1989: 168). The group was more a "technical" than a "political" alliance within the EP and consisted of the GRAEL (Green Alternative European Link, 11 members from the Greens in Germany, Belgium, and the Netherlands as well as one member from the Proletarian Party in Italy), the EFA (European Free Alliance, regionalists from Belgium and Italy), and the Danish movement against the European Community. Overall, the Rainbow Group held 20 seats in the EP.

Between 1984 and 1989 the GRAEL (the Green faction within the Rainbow Group) was heavily dominated by the German Greens, who provided the majority of the faction members. The group developed as a rather heterogeneous faction in terms of both organization and program. The GRAEL introduced the rotation principle, opened all meetings to the public, and paid the same salary to members of parliament and their staff. In programmatic terms, the GRAEL suggested building transnational strings in order to fight against transnational problems such as pollution, violation of human rights, and the arms race. Furthermore, the GRAEL considered the EP as an institution through which the Green movement could create transnational links and influence Europea-wide public consciousness rather than an institution in which Green political goals could be achieved.

In the 1989 European elections, the Greens gained considerable electoral success. Twenty-six Green Party candidates from five different countries (Germany, Belgium, France, Italy, and the Netherlands) were elected to the EP. In addition, one candidate from Portugal and one from Spain joined the Green Party federation. Thus, the German "leftist"-dominated GRAEL group was challenged by a majority of new Green members of the Parliament who belonged to the "purist" faction of Green Party thought. As a consequence, the GRAEL dissolved in summer 1989 and the Greens formed their own "Green Group in the European Parliament"(GGEP). The new group was characterized by two features. First, the Germans represented only one-third of the delegates within the GGEP. Thus, compared to the GRAEL, the Germans lost their political dominance. Second, the GGEP became a more professional political player within the EP, because the new group was no longer part of the anti-EC-oriented "Rainbow Group." For instance, Green

Party members participated more actively in EP committee work than members of other party federations (Bowler/Farrell 1992: 133). Moreover, the Greens developed a much more positive attitude toward the EP.

After the 1994 European elections, the GGEP membership composition changed considerably. On the one hand, the French Green Party representation in the EP declined to 3 percent and no seats. On the other hand, the Swedish Greens received 17 percent of the vote and gained four seats, and the number of German Green Party members within the EP increased from 8 to 12 (see table 7.1). In terms of programmatic profile and political style, the new GGEP was more similar to the GRAEL than to the GGEP of 1984–89 (Bomberg 1996: 326). First, the Mediterranean policy style diminished after the French Greens left the EP. Second, although the Germans dominated numerically, they seldom agreed among themselves on policy issues, which explained their comparably low political dominance in the new GGEP. Third, the new GGEP was constituted by individualists who devoted intensive energy to projects of their own concern, which limited the political power of the Green Party federation in general. Fourth, many of the new GGEP members were opposed to the EU integration process (in particular those from Sweden and Ireland). However, in spite of these similarities, there were also major differences between the GRAEL and the new GGEP. To begin with, there was more intensive discussion about EU policy stands like economic and monetary union, the common foreign and security policy, and the EU policy on regions. In addition, the EU position of the German Greens changed over time. While they were the strongest EU opponents in 1984, their policy merged much closer to the European integration process in 1998.

Policy Stands and Performance in the European Parliament

During the 1999 electoral campaign, the GGEP put forward a 14-point program, which was no longer determined by "left" or "pure" Green ideology. Instead, it was a rather pragmatic catalogue of requests in which the Greens asked for more transparency and more democratic control of EU institutions. In addition, the GGEP produced a common party program, which concentrated on concrete policy demands supported by all Green Parties in the EP.

The 14-item "EU institution program" reads as follows:

- Meetings of the European Council should be public.
- Commission documents should be publicly accessible in the drafting stage.
- The right of the Parliament to inspect Commission and Council documents must finally be implemented.

- Commissioners should bear individual responsibility for the tasks assigned to them and in case of incompetence have their portfolios withdrawn.
- There should be flatter hierarchies in EU institutions.
- A new code of conduct should be introduced for the Commission, in particular to ban nepotism and favoritism.
- The influence of member states on their Commissioners should be restricted.
- The new anti-fraud office, which will replace the anti-fraud unit, should be established swiftly, must have complete operational independence, and be competent for all EU institutions.
- The Parliament's right of inquiry should be extended.
- All member states should incorporate the concept of "infringement of the financial interests of the Union" into their penal law; the misuse of EU funds should be prosecuted with the same severity as misuse of national funds.
- The level required for passage of a no-confidence motion of the Parliament against the Commission should be reduced to 50 percent.
- The European Court of Justice should be given powers to impose disciplinary measures in the case of EU officials violating financial provisions.
- A public prosecutor should be appointed at European level for criminal offenses perpetrated by European institutions.
- Officials informing the Parliament or judicial authorities of irregularities within EU institutions must be protected against reprisals.

The common Green Party platform for the European elections included the following issues:

- *Peace*: The Greens pressure the EU to make its common foreign and security policy (CFSP) fully accountable to the European Parliament.
- *Environment*: The Greens demand a European environmental policy based on "four natural steps": consume resources only at a rate at which nature can replace them; allow nature the space to recover natural cycles; stop production of unnatural and persistent substances; reduce what we take from the earth's crust and disperse into the air.
- *Energy*: The Greens demand that the prices of fossil and nuclear-based energies be adjusted so as to reflect their real cost to society and the planet.
- *Human rights*: The Greens call for the adoption of a Charter of Basic Rights guaranteeing the highest standards of political, civic, economic, social, cultural, and environmental rights to all people living in the EU.
- *Employment*: The Greens request: a move to drastically reduce weekly working time through the EU; a new distribution of paid and unpaid

work between men and women; a creation of new jobs through ecological restructuring of society and an ecological reform of the tax system; a measure to develop the "third sector" of socially useful, community-oriented nonprofit activities as a source of new fully recognized, socially protected jobs; and a systematic encouragement of small and medium-sized businesses, including measures to facilitate their creation.

• *Transport*: Green policy is working toward an integrated transport system giving mobility to all EU citizens, providing maximum quality and service, and reducing damage to society and the environment.

• *Democracy*: To make the EU truly democratic, the Greens ask for a debate about a "Constitution for Europe," which sets out clearly the competences of EU institutions and guarantees the fundamental freedoms of citizens.

• *Enlargement*: The Greens consider that the EU enlargement presents ideal conditions for advancing the peaceful economic and political integration of Europe but that there are problems to be tackled if a wider EU is to work. The Greens strongly support a reform process but claim that the principles of sustainable economy must be established as binding and extended to include agriculture.

Although the programmatic performance of the GGEP was well designed by party headquarters, the performance of the Greens inside the EP was slightly contradictory for at least three reasons. Firstly, the Greens criticized the growing political power of the Commission and the Council and the subordinate role of the EP in the European decision-making process. Nevertheless, members of the GGEP participated actively in the committee system inside the EP, knowing that their political impact was rather limited. Secondly, the Greens have always asked for decentralized political structures. However, the EU policymaking structures in which GGEP members were heavily involved appeared to be predominantly centralized. Thirdly, members of the GGEP in Brussels became more and more alienated from their rank and file in the electoral districts. This contradicted the ambition and major demand of the Greens to let the grassroots participate and control the decision-making process in politics. However, in spite of this contradictory performance in the EP, the Greens have been extremely successful in the fifth, 1999, direct elections to the European parliament.

Electoral Results

The electoral results of Green Parties in Europe reads like a success story (see table 7.1). Over the course of 20 years, the Greens were able to increase their electoral support from 2.6 percent (1979) to 9.9 percent (1999). In addition,

Table 7.1 Electoral results of Green Parties, 1979–1999 (European elections, in percent and seats)

Countries	1999		1994		1989		1984		1979	
	% exp	S	% exp	S	% exp	S	% exp	S	% exp	S
Austria	9.2	2	6.8	1						
Belgium										
Agalev	12.0	2	10.7	1	12.3	1	7.1	1	2.3	0
Ecolo	22.3	3	12.9	1	16.5	2	9.8	1	5.1	0
Finland	13.4	2	7.6	1						
France	9.7	9	3.0	0	10.6	8	3.4	0	4.4	0
Germany	6.4	7	10.1	12	8.4	8	8.2	7	3.2	0
Great Britain	6.2	2	3.2	0	14.9	0	2.6	0	0.1	0
Ireland	6.7	2	7.9	2	3.7	0	0.1	0	–	–
Italy	1.8	2	3.2	3	6.2	5	–	–	–	–
Luxembourg	10.7	1	10.9	1	10.4	0	6.1	0	1.0	0
Netherlands	11.9	4	3.7	1	6.8	2	5.6	2	–	–
Sweden	9.4	2	17.2	4	–					
Average electoral result	9.9		8.1		9.9		5.3		2.6	
Total seats in European Parliament	38		27		29		11		0	

Source: Keesing's Archiv for 1999: http://www.europarl.eu.

the number of Green members in the EP increased from 11 (1984) to 38 (1999). The most successful Green Parties have been the ones in Belgium and Luxembourg. Their electoral strength increased constantly at a fairly high level over the years. The same is true for the German Greens, except for the 1999 elections, and for the Dutch Greens, except for the 1994 elections.

By comparison with the European elections of 1994, the fifth direct elections to the EP provided much better electoral results. Overall, the Greens gained 9.9 percent of the vote in 12 countries in 1999. Five years ago, they received only 8.1 percent. This result also affected the Parliamentary strength of the Greens. They were able to increase their number of seats from 27 to 38. Thus, the Green Parties currently hold 17 percent of the total seats in the EP.

Although the electoral results on aggregate level have been very impressive, not all Green Parties in Europe were as successful as in the 1994 elections. The German and the Swedish Greens, for instance, lost considerable voter support. The electoral result for the German Green Party decreased from 10.1 to 6.4 percent, and the support for the Swedish Greens went down from 17.2 to 9.4 percent. Because of the electoral law and the distribution of seats by country, the German Greens lost five seats and the Swedish Greens only two seats in the EP. The winners of the 1999 European elections were clearly the Greens in the Netherlands (+8.2), France (+5.8), Finland (+5.8), Belgium (+5.3), Great Britain (+3.0), and Austria (+2.4). In Ireland, Italy, and Luxembourg, only minor changes took place.

Green Party Delegates in the European Parliament

Single country studies have empirically verified that women and candidates from the postwar generation score high among the Green representatives in the Parliament. This is also true for the Green Party members in the recently elected EP (see table 7.2).

Among the 38 MEPs who received their mandate in the 1999 European elections, we find 20 men and 18 women. In most of the 11 countries, there is an equal distribution between male and female representatives. The exceptional cases are Great Britain and Ireland, whose four representatives are all women, and Italy, whose two members of the Parliament are both men. Overall, it seems as if the representation of women among Green MEPs is clearly higher than in other European party federations. The age of Green MEPs by country range from thirty-nine to forty-four years in Luxembourg, Austria, the Netherlands, Germany, and Great Britain; from forty-six to forty-nine years in Ireland, France, Belgium, and Finland. The oldest Green members of the Parliament are among the Swedish (fifty-five years) and the

Table 7.2 Sociodemographic profile of Green federation representatives in the European Parliament, 1999

Countries	(N)	Male (N)	Female (N)	Average Age
Austria	2	1	1	42
Belgium				
Agalev	2	1	1	49
Ecolo	3	2	1	48
Finland	2	1	1	49
France	9	5	4	48
Germany	7	3	4	43
Great Britain	2	0	2	44
Ireland	2	0	2	46
Italy	2	2	0	59
Luxembourg	1	1	0	39
Netherlands	4	3	1	42
Sweden	2	1	1	55
Total	38	20	18	47

Source: Author's calculations on the basis of information about members of the European Parliament, http://www.europarl.eu.

Italian Greens (fifty-nine years). In sum, the average age of the Greens is low (forty-seven years), compared with the other larger party federations in the EP.

The age of MEPs is also an indicator of the socializing effect of certain age cohorts. In what historical phase of postwar Europe have Green members of the Parliament been socialized? It is suggested here that five age cohorts be differentiated:

- *The early postwar generation (born before 1947)* This generation was politically socialized in the 1950s. It has a predominantly materialist-oriented value system and a stronger ideologically determined form of political behavior. Among the Green MEPs, 28 percent belong to this generation.
- *The student movement generation (born between 1947 and 1955)* This generation adopted the political ideologies of the mid-sixties, which were characterized by left-wing attitudes, solidarity, and calls for radical change of the society. Among the Green MEPs, 32 percent belong to this age cohort.
- *The local new social movement generation (born between 1956 and 1960)* This generation was still socialized by the student movement but also by newly developed citizen initiative groups, which demonstrated locally against environmental destruction and nuclear power plants. This age cohort has developed left-wing ideas mixed with ecological demands. Among the Green MEPs, 30 percent belong to this group.
- *The new social movement generation (born after 1961)* This generation was heavily influenced by the political activities of the peace, anti-nuclear power, ecology, women's rights, and other alternative movements. Their political behavior is less determined by ideological stands and more influenced by postmaterialist values such as self-determination, participatory democracy, concern with energy, environmentalism, human rights, and the Third World. Among the Green MEPs, 10 percent belong to this group.

In sum, about 60 percent of the Green MEPs in the contemporary EP belong to the student movement and the early new social movement generation, while 40 percent are socialized in other historical settings. This might explain the constant internal struggle about the political style and the different political demands among the Greens in the EP.

Green Party Voters

The literature on Green Party voters confirms that the electorate of the Greens belong to the group of well-educated citizens, who are mostly young, belong to the middle class, and are employed in the nonproductive white-collar service sectors of the economy and the state bureaucracy (Müller-Rommel 1990). The effect of youth, for instance, and to a lesser degree of education and occupation, is still a common predictor for the Green Party vote. Naturally, young age cohorts are more attracted to the policy profiles and the political style of the Greens. In addition, the young voters have not participated in many elections, and they have often not developed enduring ties with established political parties. It is therefore safe to argue that a small core Green support will constantly be derived from the age cohort of between eighteen and thirty years.

However, with the possible exception of age, the sociodemographic variables are no strong predictor for Green Party success. Two researchers, who have done empirical research on Green Party voters over the previous four European elections, also confirm that Green Party voting is not particularly narrowly defined in sociodemographic terms (Franklin/Rüdig 1995). In the absence of this explanation, who then votes for the Green Party? There is much to be said in favor of the argument that the Green voter belongs to the type of "rational voter" whose electoral choice is determined by the subjective evaluation of the parties on policy matters. Most Green voters do not only belong to the young generation; they are also highly interested and very active in politics. In addition, they are more critical about party politics than is the average voter. The voting behavior of this group is rationally guided; their electoral choice is influenced by the retrospective evaluation of competing parties' policies. Thus the policy stands of the Greens and the behavior of the Green Party elite have a substantial effect on the voting behavior of potential and actual Green Party followers.

Explaining the Electoral Success

The high electoral results of the Greens in the recent European elections came as unexpected for many political observers. Although the support for the European Greens had increased indeed from 1979 to 1999 over the last 20 years, it should not be overlooked that the electoral strength of the Greens in 1999 varied substantially across Europe. The results ranged from 1.8 percent in Italy to 22.3 percent in the French part of Belgium. Thus, there are more and less successful Green Parties in Europe. In the following analysis, we differentiate between very successful Green Parties, which have scored above 9 percent of the vote, and less successful Greens, which have received electoral results between 1 and 9 percent. According to table 7.1, Green Parties from the following countries were very successful in the 1999 European elections: Austria, Belgium, Finland, France, Luxembourg, the

Netherlands, and Sweden. Among the less successful Greens were the ones from Germany, Great Britain, Ireland, and Italy. How can we explain the variation in electoral support for these Green Parties across Europe? In the literature on Green Parties, an attempt has been made to link the rise of Green Parties to political and economic conditions in advanced industrial societies (Müller-Rommel 1998). It is widely believed that the Greens have been electorally successful as a consequence of the political environment in which they operate: It has thus been pointed out that characteristics of the political and the economic setting of a country are responsible for the electoral strength of these parties. It is therefore worth examining some factors that are typically regarded as having played a part in the electoral success of the Greens. These concern, first, the political setting: voter turnout and the presence or absence of a "cartel" form of relationship among the main parties. Second, there is an economic setting, which consists of the rate of unemployment and the rate of economic growth.

Voter turnout. Much research has been conducted to analyze the relation between voter turnout and electoral results for political parties. In a nutshell, it is argued that small parties benefit from a low voter turnout. With two exceptions, this hypothesis holds true for the Greens in the 1999 European elections. In countries like Austria, Finland, Sweden, France, and the Netherlands, we find a voter turnout below 50 percent and high electoral success for the Greens. Vice versa, in countries with a high voter turnout (above 50 percent), the Greens have been less successful (Ireland and Italy). The two exceptions are Belgium and Luxembourg, which have introduced a compulsory voting system.

"Cartelized" political system. In the mid-1990s, the concept of a "cartel" party was introduced by Katz and Mair (1995). In a nutshell, they argue that large established parties are losing sight of their electoral support base in the society; this has led to the foundation of new "challenger" parties (like the Greens) and to a decline in the electoral support for the established parties. As a reaction against this decline and in order to maintain their commanding position, the main established parties have made their relationship with state structures much closer, by means of control of the mass media, the spread of party patronage in the state bureaucracy, and the domination of both private and public companies. In other words, a cartel-type arrangement exists when all major established parties collectively cooperate with each other successfully in linking themselves with state structures and in resisting the programmatic and organizational challenges from newly founded parties. Provided that cartel parties do maximize the difference

between themselves and their electoral support base, we would expect a "cartelized system" to be conducive to the electoral success of Green Parties. According to Taggart, "cartelization" is measured by the "average number of parties in the cabinet per government per decade multiplied by the number of different parties represented in the cabinet over the same decade" (1994: 19). Using this definition, a highly cartelized system can be regarded as having existed in Finland, Belgium, the Netherlands, Austria, Sweden, and Germany. Less cartelized party systems can be found in Luxembourg, France, Ireland, Italy, and Great Britain. The hypothesis does seem to be valid for most of the cases. While Green Parties have been successful in five countries in which there was a cartel system—Finland, Belgium, the Netherlands, Austria, Sweden—they have been less successful in three party systems where there was no cartel system: Ireland, Italy, Great Britain.

Unemployment rate. Low unemployment is expected to benefit the Greens because in countries where it now exists, the new political culture develops faster than in countries with high unemployment. The unemployment rate is measured on the basis of the mean percentage of unemployment in 1998. On this basis, the Greens have been successful in all countries (except Great Britain) with a low unemployment rate (less than 10 percent)—Austria, Luxembourg, the Netherlands, and Sweden. Vice versa, the Greens received lower electoral results in three countries with a high unemployment rate—Germany, Italy, and Ireland.

Growth of national product. The growth of national product is measured by the percentage change in the GNP from one year to another. Increases over 2.7 percent are considered high. In applying Inglehart's theoretical assumptions about the relationship between economic growth and "new values/politics," it is to be expected that a high rate of growth in the GNP benefits the Greens. This assumption can be confirmed for: France, Finland, Luxembourg, and the Netherlands. For two other countries—Italy and Great Britain—the hypothesis holds true: We find a low GNP and lower electoral results for the Greens.

Given these findings, we can draw two conclusions: First, the four variables used in the above analysis help to explain the electoral performance of Green Parties in the 1999 European elections. Second, for two European countries, the political and economic variables have a strong predicting power. In the Netherlands, for example, we find low voter turnout, a highly cartelized system, a low unemployment rate, high economic growth, and a very successful Green Party. In Italy, the reverse is true. We suggest that the Greens are not very successful in Italy because of a high voter turnout, a low level of cartelization, high unemployment, and low growth of national

Table 7.3 Explanatory variables for Green Party success

Variables	Electoral results for the Greens, 1999	Low voter turnout, 1999	Highly cartelized system, 1999	Low unemployment rate, 1998	High rate of economic growth, 1998
Countries					
Belgium	17.1	–	+	–	–
Finland	13.4	+	+	–	+
Netherlands	11.9	+	+	+	+
Luxembourg	10.7	–	–	+	+
France	9.7	+	–	–	+
Sweden	9.4	+	+	+	–
Austria	9.2	+	+	+	–
Ireland	6.7	–	–	+	–
Germany	6.4	+	+	–	+
Great Britain	6.2	+	–	+	–
Italy	1.8	–	–	–	–

product. The other nine countries range somewhere in between these two extremes. The success of the Greens in Austria and Sweden can, for instance, be explained by low voter turnout, a highly cartelized system, and a low unemployment rate (see table 7.3).

The Future of the European Greens

After the 1999 European elections, the Green federation became comparably strong in terms of members and resources. Thus, there are excellent opportunity structures for playing an active role in the EP. Aside from the fact that individual Green MEPs increasingly produce reports and raise issues in a host of areas, the GGEP has also decided to increase their activities in the committee system of the house. Currently, Green MEPs participate in 18 EP committees, holding the presidency of three (Citizens, Freedom, and Rights; Women's Rights and Equal Opportunities; Culture, Youth, and Education). In addition, the Greens have delegated five MEPs in each of the following committees: Industry, External Trade, Research, and Energy; Environment, Health, and Consumer Policy; and Foreign Affairs and Human Rights. In sum, one can expect that the Greens will perform a high degree of activity inside the EP. They will be critical about the policy of the old party federations and about the democratic deficit of the entire EU structure. Thus, the GGEP will challenge the established party federations as well as the overall decision-making process within the EU.

The activities inside the Parliament have, however, only little effect on the durability of the Greens in the EP. The rise and fall of some Green Parties in western Europe are based on national rather than on European premises. The difficulty facing the Green Parties in the various European countries is that their voters show more loyalty to issues than to political parties. They assess governmental and party policy more sensibly and critically than the average voter. This can benefit the Greens under one most important condition: As long as the Green Parties do not adopt the "catchall" approach and the hierarchical organizational structure of established parties and as long as the Greens press for radical changes in national and EU governmental policy, Green Party voters will constitute a loyal core in the electorate of the Greens. This is because adopting a radical version of new political issues is beyond the reach of established parties of the left.

In most European countries, the Social Democratic rank and file and leadership divide into two groups: those with a traditional left-wing outlook who are concerned with the strength and stability of the economy and democracy in traditional industrial society (old left), and those with a new-politics orientation who focus on the quality and the nature of the economy and the extent of democracy (new left). The "New Left" competes with the followers of Green Parties, while the "Old Left" still clings to the old cleavage dimension. The established left-wing parties are therefore trapped between two cultures. Only a minority of the electorate is on the new-politics side, while the majority of the Social Democratic rank and file are located in the center of the political spectrum. Whatever the party might be able to gain on the New Left side, it risks losing on the Old Left of the party's spectrum. Consequently, the only viable strategy for the Social Democrats and Socialists in western Europe is to attempt to conciliate old politics (in order to integrate the majority of the left-wing voters) with a moderate version of the new politics (in order to attract the Green Party voters). In such a context, the Green Parties' electoral performance depends on their adopting radical issue positions that conventional left-wing parties are not able to advocate in a full-blooded form. It seems therefore reasonable to predict that Green Parties in Europe are here to stay as long as new-politics issues remain on the political agenda and are not adopted in a radical form by any other major party.

References

Bowler, Shaun, and David Farrell (1992), " The Greens at the European Level," in *Environmental Politics*, vol. 1, pp. 132–137.

Bomberg, Elizabeth (1996), "Greens in the European Parliament," in *Environmental Politics*, vol. 5, pp. 324–331.

Bomberg, Elizabeth (1998), Green Parties and Politics in the European Union. London: Routledge.

Buck, Karl (1989), "Europe: The Greens and the Rainbow Group in the European Parliament," in Ferdinand Müller-Rommel (ed.), New Politics in Western Europe. Boulder: Westview Press.

Franklin, Mark, and Wolfgang Rüdig (1995), "On the Durability of Green Politics: Evidence from the 1989 European Election Study," in Comparative Political Studies, vol. 28, pp. 409–439.

Katz, Richard, and Peter Mair (1995), "Changing Models of Party Organization and Party Democracy: The Emergence of the Cartel Party," in Party Politics, vol. 1, 5–28.

Müller-Rommel, Ferdinand (1985), "Das grün-alternative Parteienbündnis im Europäischen Parlament," in Zeitschrift für Parlamentsfragen, vol. 16, pp. 391–404.

Müller-Rommel, Ferdinand (1990), "New Political Movements and 'New Politics' Parties in Western Europe," in Russell Dalton and Manfred Küchler (eds.), Challenging the Political Order. Oxford: Polity Press, pp. 209–231.

Müller-Rommel, Ferdinand (1993), Grüne Parteien in Westeuropa. Opladen: Westdeutscher Verlag.

Taggart, Paul (1994), "Riding the Wave: New Populist Parties in Western Europe," paper for the annual meeting of the European Consortium for Political Research (ECPR workshops), Madrid.

CHAPTER 8

Ethnoregionalist Parties and European Integration at the 1999 European Elections

Lieven De Winter

Introduction

The analysis of the relationship between European integration and the success of ethnoregionalist parties in Europe has received up until now very little scholarly attention. This lacuna is partially due to the lack of comparative research treating this group of parties as a genuine European party family. This neglect is reinforced by the feeble representation of these parties in terms of seats in the European Parliament, which is in the first place due to the relatively small size of the regions in which they compete for votes (De Winter and Türsan, 1998).[1] Finally, ethnicity in modern societies is often seen as a "vestigial phenomenon" (Esman, 1977: 371) and particularly within the context of an integrated Europe and the gradual functional disappearance of the old nation-states.

However, for this party family, the link is highly significant, as European integration radically modifies—and generally in a positive way—the structure of opportunities of regions and ethnoregionalist parties.

Definition of Ethnoregionalist Parties and their Relationship with European Integration

From the plethora of labels and definitions that one finds in the literature, we prefer to use the term "ethnoregionalist," which we define on the basis of the two common denominators that unite these parties: (1) a subnational territorial division and (2) a population that the ethnoregionalist party assumes to constitute a category that is culturally distinct and has an exclusive group identity. Their programmatically most defining characteristic is their demand for empowerment of the ethnoregional collectivity. This demand thus calls for the reorganization of the power structure of the national political system, for a certain degree of self-government for the region.

Viewed from the perspective of a process of transfers of competences from the national level to a higher level (and in federal and regional states, also competences that belong to the regions and federal states), European integration undoubtedly constitutes an amplification of the "democratic deficit" defined in terms of distance between decision makers and the beneficiaries of public policies.

The Implausible Beneficial Effect of European Integration on Ethnoregionalist Parties

Historically, the founding fathers of European integration were strongly anti-nationalist, as nationalism in all its forms was considered to be the main cause of the "European civil war" that ran from 1933 to 1945 (Lynch, 1998: 179). The European Coal and Steel Community intended to put the basic industry necessary for (French and German) rearmament under supra-national control. The creation of Euratom was inspired by similar objectives. European integration essentially aims at diminishing the sovereignty of the national state, and therefore the ambition of regions to become new states in the nineteenth century sense may appear anachronistic.

European integration represents a process of centralization of the decision-making process. Initially restricted to the economic activities linked to rearmament, this centralization has spread into other sectors of economic and monetary policymaking, recently starting to include aspects of foreign policy, national defense, and internal security. This process should logically widen the gap between regional populations and beneficiaries of public policies on the one hand, and, on the other hand, the main decision-making centers regarding policies relevant to the populations that ethnoregional parties want to empower.

In addition, European integration and economic globalization may exacerbate territorial disparities, further peripheralize marginal regions, and reinforce the stronger regions (in the "golden triangle" or the "blue banana"). The accentuation of these regional disparities and the occurrence of asymmetrical shocks cannot anymore be countered by public intervention: Regions cannot devalue their currencies in order to restore competitiveness, EU competition policy prohibits state subsidies to ailing industrial sectors concentrated in certain regions, and the use of deficit spending and the manipulation of interest rates have been severely restricted by the convergence criteria of the

Maastricht Treaty (De Grauwe, 2000). In spite of the rhetorical and genuine efforts of EU regional and structural funds policy, the budget for reducing regional disparities remains on the average below 2 percent of GDP of EU member states. In addition, economic disparities are reinforced by political ones: In the most marginal regions—in which EU regional policies constitute a considerable part of the gross regional product—political ethnoregional mobilization is generally weak, while ethnoregionalist parties tend to flourish most strongly in the richer regions (Fearon and Van Houten, 1998).

European integration poses also a series of political and constitutional challenges to the regions, as EU competences also affect those policy domains devoluted to regions in federal states (hence Article 146 of the Maastricht Treaty allowing regional ministers to represent their country in the Council of Ministers for matters decided in their country on the regional level). In addition, the most important channel of access of regions to the EU decision-making process is their national government. The better that regional interests are integrated in the national policymaking system, the better they will be looked after in Brussels (Keating, 1998: 166).

Thus European integration implies the centralization of decision making at the European level for a large number of policy sectors. It comes as no surprise that at the outset of this process, many ethnoregionalist parties opposed European integration, given the widening of the gap between regional populations and supranational decision-making centers (Lynch, 1998). However, as Europeanization of decision making expanded and accelerated, paradoxically their Euroskeptical position gradually evolved into a strong Europhilic stand.

The Feeble European Integration of Ethnoregionalist Parties as a Political Family

At this moment, the most concrete form of European integration of the family of ethnoregionalist parties is the Democratic Party of the Peoples of Europe—European Free Alliance, which rallies most regionalist and autonomist parties in Europe (see table 8.1).

The cooperation between ethnoregionalist parties in the European Parliament underwent different phases that well illustrate the problems with founding a genuine trans-European party (Lynch, 1998). After the 1979 elections, the members of the DPPE-EFA formed a technical group with some extreme-left parties. In the following legislature, this group was enlarged with the Greens (the "Rainbow" group). Only in the 1989–1994 legislature was a homogenous ethnoregionalist parliamentary group formed. From 1994 to 1999, the remaining three MEPs representing DPPE-EFA parties joined the French and Italian radicals (the European Radical Alliance

Table 8.1 The Democratic Party of the Peoples of Europe—European Free Alliance as a representative of a European political family, October 1999

Members of the DPPE-EFA	Ethnoregionalist parties not affiliated with the DPPE-EFA
Bloque Nacionalista Galego (1)	Convergencia Democratica di Catalunya (2, ELDR)
Eusko Alkartasuna (1)	Coalicion Canaria (1, ELDR)
Esquerra Republicana de Catalunya	Euskal Herritarrok (1, nonaffiliated)
Fryske Nasjonale Partij	Lega Nord (4, technical group and nonaffiliated)
Mouvement Région Savoie	Svenska Folkpartiet (1, ELDR)
Partei Deutschsprachiger Belgier	Südtiroler Volkspartei (1, EPP)
Partido Sardo d'Azione	
Partit Occitan	
Plaid Cymru (2)	
Scottish National Party (2)	
Slovenska Skupnost	
Union Démocratique Bretonne	
Union für Südtirol	
Union du Peuple Alsacien	
Union Valdôtaine	
Unione di u Populu Corsu	
Unitat Catalana	
Volksunie (2)	
Partido Nacionalista Vasco (1)	
Partido Andalucista (1)	

Source: DPPE-EFA; Green-EFA group.
*In brackets the number of MEPs, and for non-members of the DPPE-EFA, the parliamentary groups to which they belong.

group), while in the current parliament, the ten MEPs of the DPPE-EFA have joined the Green group (which now counts a total of 48 MEPs). In short, apart from one legislature, the DPPE-EFA representation in the European Parliament has never been sufficiently strong to form a genuine ethnoregionalist parliamentary group. In addition, in most cases, the ethnoregionalist MEPs constituted a small minority in the group they joined.[2] Thus, given the low number of MEPs, most parties of the DPPE-EFA have been excluded from the party integration opportunities that the Parliamentary arena offers (Hix and Lord, 1997).

On the other hand, several important ethnoregionalist parties prefer to adhere to other Parliamentary groups, which further undermines the representative character of the DPPE-EFA. After the 1999 elections, there are in fact ten MEPs belonging to ethnoregionalist parties that are not members of the DPPE-EFA, thus as many as currently belong to the DPPE-EFA parties. The most notorious outsiders are the Lega Nord, the CiU, the SFP, the StVP, and the HB. Note however that the current numerical balance between DPPE-EFA members and nonmembers is unprecedented, indicating an evolution in favor of the DPPE-EFA. The preceding Parliament for example counted only 3 ethnoregionalist MEPs belonging to the DPPE-EFA, against 11 nonmembers.

In short, at the level of the European Parliament, the DPPE-EFA is neither inclusive as a transnational party nor predominant in the definition of the political outlook of the Parliamentary group to which it belongs. This constitutes a serious handicap in comparison with other European party families that are clearly more inclusive and homogenous, and thus more representative of the ideological tendency they articulate.

Europe as an Opportunity Structure Favorable to Ethnoregionalist Parties: The Official Position of the European Free Alliance toward the EU

In spite of all the aspects of European integration that undermine the interests of the target publics of ethnoregionalist parties, these parties have adopted—at least since the first direct elections to the European Parliament—favorable attitudes toward European integration.

The objectives of the DPPE-EFA with regard to European integration are the following: the construction of a European Union of free peoples living in solidarity, founded on the principle of subsidiarity; the defense of human rights and rights of peoples, and in particular the right to self-determination; the protection of the environment and a sustainable development; the construction of a just society based on solidarity and progressive policies, social cohesion, and equal opportunity for all citizens; and the participation in European politics of parties that, due to their dimension, the electoral system, or the size of the territory they represent, have been excluded from representation.

The principles and demands formulated in the EFA's manifesto prepared for the 1996 Intergovernmental Conference offer a more precise image of this party's pro-integrationist and federalist stands: more supranationality in EU decision making, a social Europe parallel to the European Monetary Union (including social convergence criteria); a common defense and foreign policy beyond simple intergovernmentalism; enlargement of the EU, but in concentric circles, allowing to associate with the center states (of the EMU) certain countries that cannot fulfill immediately all criteria, cooperating on the basis of federalism; bicameralism in the form of a directly elected Senate of the Peoples and Regions, a sort of combination of the Committee of the Regions and the Council of the Regions; an elected European government by a majority in the two chambers; the right of initiative of European, national, and regional parliaments as well as the Committee of the Regions; the creation of consultative committees of MEPs, MNPs, and regional MPs to improve cooperation among different parliaments; in the short term, the representation of the regions in the Council of Ministers for the matters that are devoluted in their country to the regional level, splitting the votes of the states in the Council of Ministers among regions; assuring the direct access of regional authorities and the Committee of the Regions to the European Court of Justice; the use of the unanimity rule for cultural and citizenship matters; and installation of an alarm-bell mechanism available to regional assemblies that want to block European decisions that hurt vital interests of the region.

The ethnoregionalist parties that are not members of the DPPE-EFA but of one of the main Europarties are equally in favor of integration, and in general more strongly so than the official line of their host party. However, even if almost all ethnoregionalist parties are now defending European integration, large divergences between these parties exist concerning the model of further integration the EU should pursue. Not all ethnoregionalist parties are Euroenthusiasts. Some, like the Lega Nord, the BNG, and the SNP, are less in favor of the creation of a supranational Europe and promote rather an intergovernmental or confederal model in which their regions would constitute a proper nation-state. In addition, even when many parties evoke a Europe of the Regions, a federal model with only two levels (Europe and the regions) is not explicitly advanced. So the end of the state to which they belong is not (yet) announced, though nearly all ethnoregionalist parties do demand a stronger presence of their region in the delegations representing their countries within EU institutions.

Concerning the division of competences among Europe, the state, and the region, the positions of ethnoregionalist parties vary between two poles: on the one hand, the desire to transfer massively entire policy sectors to the European level (generally competences that at this moment still are exercised by the states, but whose scope is too transnational or comprehensive to be efficiently exercised at the regional level, like defense and foreign policy, monetary and fiscal policy, large-scale public works, the environment, security, etc., but also employment policy), and on the other hand, the expansion of the competences of the regions at the cost of the EU, a position rarely defended, in spite of the fact that in most electoral manifestos and ideological charts, subsidiarity is evoked as the sacred basic principle (and, not surprisingly, is interpreted only in its bottom-up version).

More recently European integration has boosted the hopes of most ethnoregionalist parties. Several factors can explain the shift of attitudes in favor of European integration.

The Failing Argument against *Kleinstaaterei*

European integration, together with other forms of international integration and cooperation, the enlargement of scale of traditional state functions, and the globalization of economic activity have weakened the classical arguments against *Kleinstaaterei* (Hobsbawm, 1992: 31).

First, the creation of customs and a monetary union has brought economies of scale within reach even for small producing countries. Their products have gained access to large markets, while international producers have also found small countries attractive for investments, as long as they are part of an open economy and supportive of business interests by specific development programs. Second, for those regions that aspire to independence, the introduction of the euro solves the problem of monetary transaction costs that a new independent region-state would face, creating its own currency and defending it on international markets, during a transition period most likely characterized by turmoil and disorder typical for separatist processes. Finally, the success of NATO indicates that in the nuclear age, even former European superpowers like France and the U.K. have to appeal for their defense to a larger international cooperation in order to perform this function, essential to the classical nation-state. In a certain sense, large countries have also become *Kleinstaaten,* not capable anymore of performing classical state functions in a satisfactory way.

Hence, the new international institutional context has on the one hand permitted western European states to survive and prosper (Milward, 1992), and not only small ones like the Benelux countries, Denmark, and Ireland, but by now also the larger ones; while on the other hand, it has reduced the economic and military costs of the option of "independence within Europe."

The Weakening of the State from Above and Below

The transfers of competences from the national level to the EU and other international organizations like NATO have gradually weakened the national state "from above." At the same time, in most countries with ethnoregional party mobilization, the unitary state was weakened "from below" through the process of asymmetric federalization (Belgium, Spain), regionalization or devolution (France, Italy, U.K.).

The launching of the principle of subsidiarity, which was originally intended only to regulate the division of competences between the EU and the national states, has been seized upon by the Committee of the Regions and other regional platforms as applicable (and by now justifiably) also to the division of competences between regions on the one hand and the EU as well as the national states on the other hand.

Hence, in relative terms, the regional level gained importance as a policy-making level vis-à-vis the state level.

The Recognition of the Regions by
Specific EU Regional Programs

Europe has massively invested in regional policies that accord to the poorest regions a substantial economic support, which they otherwise would not have obtained from the state to which they belong. These programs have reinforced the regions as a relevant decision-making level, even in states where regions did not exist or lacked significant competences. The EU regional policies require solid partners at the regional level, in the phase of policy preparation as well as its implementation. Apart from political partners (like regional and local executives), the European Commission invites also interest groups to participate in these phases. The regions are thus forced to constitute themselves as competent actors to represent their regional interests in Brussels, and this through a multitude of channels of access to EU decision making (Commission, Council, Parliament, Committee of the Regions, and other forms of cooperation between regions and cities, lobbies, etc.). This decision-making model facilitates or reinforces policy networks among political, socioeconomic, and administrative actors, at the regional level as well as at the (inter)communal, transregional, and transborder level (Keating, 1998: 176).[3]

At the symbolic level, this decision-making process has projected regions and regional politicians into the European arena, presenting them as important participants in the EU policy process. They allow regional politicians to take credit for attractive EU subsidies, even those that would have come in any case simply by the working of the relevant eligibility rules (Keating, 1998: 170). In the regions in which an ethnoregionalist party is hegemonic or predominant, the leaders of these parties cash in the symbolic dividends, while in the regions with feeble ethnoregionalist mobilization, statewide parties take most credit.

When a policy-making level gains in importance in terms of issue salience and prestige, political and socioeconomic actors will gradually pay more attention to it, in terms of political personnel and campaign resources, and will tailor their programs to respond better to socioeconomic as well as identity demands of the region.[4]

If we follow the logic of Newman (1994: 41), who claims that the creation of regional policies by European states in the 1950s and 1960s, rather than their policy-centralizing tendencies, was at the basis of the breakthrough of ethnoregionalist parties in the 1960s and 1970s, we can expect that the EU recognition of the regions through its cohesion policymaking process will have the same effect. In fact, in both cases the objective was the same: to reduce unequal regional development. The states tried to enhance the fair distribution of the benefits of the expanding welfare state during the postwar economic recovery and boom, while the structural and cohesion policy of the EU is an attempt to help the weaker regions of Europe to bridge the development gap with the richer regions, whose fortunes would be further enhanced by the realization of the internal market. But contrary to the centralized and technocratic regional development programs of the national states in the 1960s, the EU has from the beginning tried to incorporate regional actors. The principle of subsidiarity and the necessity of lobbying are additional elements that may give regional cohesion policies a stronger supportive impact on the development of ethnoregionalist parties than the preceding national waves of regional policies had.

European Demonstration Effects

Transnational demonstration effect refers to the impact of the success of ethnoregionalist/nationalist movements in one state on the development of self-confidence of similar movements in other states and eventually to the founding of a genuine party. At the European level, these demonstration effects are strongest if contacts among such movements are institutionalized, as is for instance the case of the European Free Alliance. The European Parliament offers to the ethnoregionalist parties an arena for organizing meetings, cooperation, and the elaboration and articulation of a common program. The Committee of the Regions offers similar opportunities. Such cross-national interregional networks can provide weaker movements with logistics, programmatic support, and political status and prestige and, last but not least, boost their morale.

European Elections as a Springboard for the Mobilization of Ethnoregionalist Parties

According to Lynch (1996), most ethnoregionalist parties obtain generally better results at the European elections than at the parliamentary elections in their country. Several factors can explain this better performance at the European level:

(1) Among new parties, ethnoregionalist parties are the least vulnerable to the sanctions imposed on third parties by a majoritarian electoral system. Given that by definition the electorate of an ethnoregionalist party is territorially strongly concentrated and therefore controls more often a relative majority in a number of single-member constituencies, the electoral system in use at the European elections are in most cases less disadvantageous to ethnoregionalist parties than the one used for general elections.

First, at the European elections, the average size of constituencies (in terms of seats to be conquered) is in all countries with significant ethnoregionalist participation larger than for general elections. This larger size should reduce strategic voting and enhance the proportionality of the allocation of seats

(Lijphart, 1994: 98–100). Second, France, Great Britain, and Italy (for three-quarters of the seats) are the only European countries using a plurality system for their general elections. But for the European elections, these three countries have adopted a multimember system (Great Britain for the first time in 1999).

These differences in electoral rules (larger constituencies and generalization of the use of proportional representation) should work in favor of ethnoregionalist parties: With an equal number of votes at the European and general elections, they should receive a number of seats in Strasbourg more in proportion to their electoral support. In fact, if one compares the degree of disproportionality (the percentage in terms of votes minus the percentage in terms of seats) at the 1999 European elections to that of the preceding general elections, the degree of disproportionality is, for all ethnoregionalist parties except the SFP, lower at the European elections—thus also for hegemonic or predominant ethnoregionalist parties, which we could expect to profit under a more disproportional system.

(2) The European elections facilitate the formation of cartels of ethnoregionalist parties of different regions,[5] which for the general elections necessarily run separately. At the 1999 elections, one found two significant ethnoregionalist cartels (in terms of seats): the Coalicion Europa (comprising the Coalicion Canaria, the Partito Andalucista, the Union Valenciana, and the Partito Aragonés) and the Coalicion Nacionalista Europa de los Pueblos (comprising the Partido Nacionalista Vasco, Eusko Alkartasuna, Esquerra Republicana de Catalunya, and Unio Mallorquina).[6] Given the nationwide constituency used for these elections in Spain, the formation of such electoral "coalitions" was the most efficient way to fight the dilution of votes for these parties.

(3) Turnout at the European elections is noticeably lower than for general elections. Ethnoregionalist parties seem to profit from this lower turnout, as their voters participate more eagerly at the European elections than at those of their statewide competitors. The ethnoregionalist electorate in fact displays a number of sociopolitical characteristics that enhance participation: Young voters, men, a high level of educational attainment, and (new) middle-class origins are—or were[7]—overrepresented (Blondel, Sinnott, and Svensson, 1998: 200–215; De Winter, 1998; Ackaert and De Winter, 1993, ICPS, 1998). The typical voter of these parties corresponds to the "new voter" of the area of "citizen politics" or "new politics," the "dealigned protest voter" who is most likely to abandon traditional parties.

(4) As European elections are often considered "second-order elections,"[8] in which citizens' votes are determined more by parties' national programs and governmental performance than by their stands on European issues, parties that do not govern at the national level (which is the case for most ethnoregionalist parties) tend to obtain a better result at the European than

at the national elections. This is due to the fact that European elections offer the opportunity for dissatisfied followers of governing parties to manifest their disagreement with government policies,[9] given the fact that the consequences of their voting rebellion are less decisive, as the European elections do not affect the formation of national executives. Thus, in European elections, the electorate of traditional parties is more open to alternative views and messages of the smaller opposition parties. The transfer of votes provoked by the malaise vis-à-vis governmental parties—illustrated at the last European elections by the flagrant defeat of the Social Democratic Parties in most of the 13 countries in which this family participates in the national government (see the chapter of Gérard Grunberg in this volume)—works partially in favor of ethnoregionalist parties.

(5) Finally, for ethnoregionalist parties, campaigning for European elections is comparatively cheaper than for general elections. As mentioned above, the number of constituencies is smaller, often coinciding with the entire region (or the entire country). These parties have to recruit less candidates and can focus their campaign resources on a few top leaders. In addition, the European Parliament allocates generous subsidies to parties that obtained seats in the outgoing Parliament. This can neutralize or attenuate certain imbalances provoked by the lack of a system of public party finance at the national level.

The Performance of Ethnoregionalist Parties at the 1999 European Elections: A Spatial, Longitudinal, and Multilevel Comparison

Did these five factors that in theory favor the electoral performance of ethnoregionalist parties also play a part at the 1999 European elections? We will examine the performances of the most significant ethnoregionalist parties (those that obtained in 1999 at least 5 percent of the votes in their region) at the European and general elections since 1979 (or since the adhesion of their country to the EU) and compare their scores measured at the level of the region, except for cases in which the territory of the ethnoregionalist party is not well defined and thus for which electoral performance can be measured only at the national level. These exceptions include the Lega Nord, which in its latest "constitution" includes all the regions north of Rome (!), and the SFP, given the relative dispersion of the Swedish minority in Finland.

Lynch's hypothesis[10] of the systematically superior performance of ethnoregionalist parties at the European elections in comparison with the preceding general elections is:

• reconfirmed for the entire period for Plaid Cymru and the SNP (29.6 percent and 27.2 percent of the regional vote at the 1999 European elections, having since 1979 always obtained better scores at the European elections, with a difference in 1999 of +19.7 percent and +5.2 percent respectively), the Svenska Folkspartiet[11] (6.8 percent, or a difference of +1.7 percent at the overall level of Finland), the Esquerra Republicana de Catalunya (6.2, or +2.0 at the level of Catalonia), the Euskal Herritarrok (ex-Herri Batasuna) (19.9, or +7.2 in the Basque country), the Bloque Nacionalista Galego (22.2, or +9.6 in Galicia), and the Unio Mallorquina (5.1, or +3.5);

• confirmed for the first time for the Volksunie (12.2 percent in the Flemish electorate or an increase of 2.6 percent), which in all the preceding European elections obtained a score clearly inferior to the general elections. This tendency was probably due to the general decline of this party, which is losing its reason for existence given the gradual (con)federalization of the country (De Winter, 1998). The gain at the 1999 European elections (which coincided with the general and regional elections) is undeniably due to the fact that its young, dynamic leader was heading the European list. His populist style allowed him to capture a part of the growing protest vote in a system whose legitimacy has been strongly eroded by the scandals of the 1990s.

• disconfirmed, again, for the Union Valdôtaine (45.9 percent), which since the 1979 European elections has obtained a score clearly inferior to that of the general elections (a difference of −2.7 percent in 1999, which is however the smallest since 1979, being at 18.8 percent in 1989). The hypothesis is equally disconfirmed for the Lega Nord for the European elections of 1994[12] and 1999 (4.5 percent at the national level, or −5.6 percent in comparison to its score at the 1996 general elections), but this is largely due to the instability of electoral alliances among political parties following the crisis of the old partitocratic regime combined with the introduction of a predominantly majoritarian electoral system. The bad score[13] of 1999 illustrates the gradual but inevitable decline of this party in the context of the recomposition of a party system after the implosion of the old center parties due to the "clean hands" operation;

• disconfirmed, for the first time (but only slightly), for the Partido Nacionalista Vasco-Eusko Alkartasuna (with 34.5 percent, or only 0.5 percent less than their aggregated scores for the 1996 general elections).[14] At the 1987, 1989, and 1994 European elections, the combined score of these two moderate nationalist Basque parties was always higher than their results at the general elections;

• neither confirmed nor disconfirmed for Convergencia i Union, which obtained only 0.2 percent more at the 1999 European elections (29.8 percent) than at the general elections of 1996. Of the three other European elections held in Spain, the score of the CiU was higher than the one obtained at the general elections once (1989), and below it twice (1987 and 1994). The hypothesis is neither confirmed nor disconfirmed for the Südtiroler Volkspartei: With 56 percent it, doubles in Süd-Tirol province its score at the 1996 general elections.[15] Yet, previously, its score at the European elections was usually below its score at the general elections (except in 1984).

If one aggregates results at the national level, one can conclude that in comparison with those of the 1994 European elections, ethnoregionalist parties improved their scores at the 1999 elections in Belgium[16] (7.2 percent, or an increase of 2.8 percent), in Finland (6.6, or +1.0), in Great Britain (4.5, or +1.3), and in Spain (14.0, or +2.7). Only in Italy was there a decline (5.7, or −1.5).

The progress of ethnoregionalist parties is even more striking in terms of gains in seats than in votes: The new European Parliament counts 20 ethnoregionalist MEPs, against 14 in the previous legislature.

Conclusion

European integration radically modifies the structure of opportunities for ethnoregionalist parties. On the negative side, there is the widening of the gap between regional populations that ethnoregionalist parties want to empower and the main decision-making centers regarding policies relevant to these populations; the increase of regional disparities; and the weakening of regional competences as the EU also gains competence in policy domains devoluted to regions in federal or regionalized states. As far as the representation of these parties at the level of the European Parliament is concerned, the DPPE-EFA is neither inclusive as a transnational party nor predominant in the definition of the political outlook of the Parliamentary group to which it belongs. This constitutes a serious handicap in comparison with other European party families that are clearly more inclusive and homogenous, and thus more representative of the ideological tendency they articulate.

In spite of all these negative aspects of European integration for ethnoregionalist parties and their target publics, these parties have adopted—at least since the first direct elections to the European Parliament—favorable attitudes toward European integration, given the fact that this integration process has weakened the classical arguments against *Kleinstaaterei;* it has gradually weakened the national state "from above"; its regional policies have reinforced the recognition of the regions as a relevant policymaking level; it has triggered demonstration effects among these parties at the European level; and, finally, European elections serve for a variety of reasons as a springboard for the mobilization of ethnoregionalist electorates.

In fact, the 1999 European elections confirm that most ethnoregionalist parties obtain better results at European than at general elections.

Notes

1. These parties exist however in nearly all EU member states. In several of them, they are party-system relevant, either in terms of the size of their electoral support or in terms of participation or blackmail potential in the formation of government coalitions among statewide parties. Additionally, in Belgium and Italy, ethnoregionalist parties have contributed to the demise of existing party systems, and in the long run may contribute to the breakup of these countries. They seem capable, together with the Greens and the populist right-wing parties, of cashing in on the recent increase in public dissatisfaction toward the established political elites and traditional political parties.

2. Apart from the Rainbow group in the 1989–1994 period, which comprised 9 EFA MEPs out of a total of 15.

3. With the insertion of regional political and civic actors into the European decision-making process, and the extension of the subsidiarity principle to the regions, the regional level also contributes to the legitimization of the integration process and therefore tends to reduce the democratic deficit.

4. Cf. the evolution of the traditional parties in Belgium (De Winter and Dumont, 1999) and Spain (Colomé, 1989).

5. At the 1989 European elections, the Corsican (Unione di u Populu Corsu) and Brussels (Front Démocratique des Francophones) ethnoregionalist parties formed an electoral alliance with Green lists (Olivesi, 1998). At the 1984 European elections, the Union Valdôtaine formed a cartel with the Partito Sarde d'Azione.

6. Another cartel, the Union de Regiones (comprising regionalist formations of Almeria, Baleares, Castilla-La Mancha, Canarias, and Madrid (!)) obtained only 0.04 percent of the votes.

7. Yet, after one or two decades, most of the specific sociodemographic features of the ethnoregionalist electorates tend to fade away (De Winter, 1998: 232–234).

8. Reif, 1985; van Der Eijk and Franklin, 1996. For a recent empirical critique of this model, see Blondel, Sinnott, and Svensson, 1998.

9. At the general elections, on average 62 percent of the incumbent governing parties lose votes at the next general elections (Müller and Strøm, 1997: 744).

10. Formulated by Lynch on the basis of an inquiry that included only four ethnoregionalist parties.

11. Since the first European elections in Finland (1996).

12. At the 1989 European elections, the Lega obtained 1.8 percent of the Italian vote, against 0.5 percent at the 1987 general elections. The bad result in June at the 1994 European elections was due to the appearance of "deus ex machina" Berlusconi and his governmental team, which still enjoyed their state of grace after the shock of the general elections of April 21, 1994.

13. At the campaign for the 1999 European elections, the Lega strongly criticized the current model of European integration (the 'Eurocracy of Brussels' and the new 'European super-state'), which gave the impression that the Lega was the only anti-European party (with Rifondazione Comunista), even when its manifesto for these elections was quite in favor of integration.

14. At the 1999 European elections, these two parties presented a common list. The EA is a split-off (1986) of the PNV (Acha Ugarte and Pérez-Nievas, 1998).

15. Note that this 1996 score, the weakest in its entire history, was due on the one hand, to internal dissidence that led to the creation of the Union für Süd-Tirol and, on the other hand, to the general turmoil of the Italian party system.

16. The Front Démocratique des Francophones has constituted since 1993 a federation with the Francophone Liberals (PRL). Since then, this federation has presented common lists at the European, general, and regional elections, but so far not at the local elections. At the last local elections, the FDF lists obtained 13.2 percent of the votes in the Brussels communes, or 0.9 percent of the Belgian voters (Bulens and Van Dyck, 1998). The Belgian ethnoregionalist vote should therefore be marginally increased. One should remember that the parties represented in the federal parliament are all regional parties. Since 1978 (the split-up of the last traditional party), there have been two distinct party systems: a Flemish and a Walloon. In the Flemish constituencies, only Flemish parties compete for votes, and they do not present any lists in the Walloon constituencies (and vice versa). Only in the Brussels-Halle-Vilvoorde constituency do these two party systems overlap, and Flemish as well as Francophone parties compete for the same set of voters (De Winter and Dumont, 1999).

References

Acha Ugarte, B., and S. Pérez-Nievas, "Moderate Nationalist Parties in the Basque Country: Partido Nacionalista Vasco and Eusko Alkartasuna," in L. de Winter and H. Türsan (eds.), *Regionalist Parties in Western Europe*. London: Routledge, 1998, pp. 87–104.

Ackaert, J., and L. de Winter, "De afwezigen hebben andermaal ongelijk. De stemverzaking in Vlaanderen op 24 november 1991," in M. Swyngedouw, J. Billiet, A. Carton, R. Beerten, *Kiezen is verliezen. Onderzoek naar de politieke opvattingen van Vlamingen*. Leuven: ACCO, 1993, pp. 67–82.

Blondel, J., R. Sinnott, and P. Svensson, *People and Parliament in the European Union. Participation, Democracy, and Legitimacy*. Oxford: Clarendon Press, 1998.

Bulens, J., and R. Van Dyck, "Regionalist Parties in French-Speaking Belgium: The Rassemblement Wallon and the Front Démocratique des Francophones," in L. de Winter and H. Türsan (eds.), *Regionalist Parties in Western Europe*. London: Routledge, 1998, pp. 51–69.

Colomé, G., *El Partit dels Socialistes de Catalunya. Estructura, funcionament i electorat 1978–1984*. Barcelona: Barcelona Edicions 62, 1989.

Dahl, R. *Democracy and its Critics*. New Haven: Yale University Press, 1989.

De Grauwe, P. (2000), *Monetary Policy in Euroland*. Paper prepared for the Vlaams Wetenschappelijk Economisch Congres.

De Winter, L., "The Volksunie and the Dilemma between Policy Success and Electoral Survival in Flanders," in de Winter and Türsan (eds.), *op. cit.*, pp. 28–50.

De Winter, L., "Conclusion: A Comparative Analysis of the Electoral Office and Policy Success of Ethnoregionalist Parties," in de Winter and Türsan, *op. cit.*, pp. 204–247.

De Winter, L. and P. Dumont, "Belgium: Party System(s) on the Eve of Disintegration?" in D. Broughton and M. Donovan (eds.), *Changing Party Systems in Western Europe.* London and New York: Pinter, 1999, pp. 183–206.

De Winter, L. and H. Türsan (eds.), *Regionalist Parties in Western Europe.* London: Routledge, 1998, p. 259.

Esman, M. J. (1977), "Perspectives on Ethnic Conflict in Industrialized Societies," in M. J. Esman (ed.), *Ethnic Conflict in the Western World.* Ithaca: Cornell University Press.

Fearon, J. and P. van Houten (1998), *The Politicization of Cultural and Economic Differences. A Return to the Theory of Regional Autonomy Movements.* Paper prepared for delivery at the 1998 annual meeting of the American Political Science Association, Boston, September 3–6, 1998.

Hix, S. and C. Lord, *Political Parties in the European Union.* London: Macmillan, 1997.

Hobsbawn, E. J. (1992), *Nations and Nationalism since 1780. Programme, Myth, Reality.* Cambridge: Canto.

ICPS, 1997, *Sondeig d'Opinio Catalunya,* Institut de Ciencics Politiques i Socials, Barcelona, 1998.

Keating, M. (1998), *The New Regionalism in Western Europe. Territorial Restructuring and Political Change.* Cheltenham: Edward Elgar.

Lijphart, A. (1994), *Electoral Systems and Party Systems. A Study of Twenty-Seven Democracies 1945–1990.* Oxford: Oxford University Press.

Lynch, P. (1996), *Minority Nationalism and European Integration.* Cardiff: University of Wales Press.

Lynch, P. (1998), "Co-operation between Regionalist Parties at the Level of the European Union: The European Free Alliance," in de Winter and Türsan (eds.), *op. cit.*, pp. 190–203.

Milward, A. (1992), *The European Rescue of the Nation-State.* London: Routledge.

Müller, W. C., and K. Strøm, "Schluss: Koalitionsregierungen und die Praxis des Regierens in Westeuropa," in Müller and Strøm (eds.), *Koalitionsregierungen in Westeuropa. Bildung, Arbeitsweise und Beendigung.* Wien: Zentrums für angewandte Politikforschung, 1997.

Newman, S. (1994), "Ethnoregional Parties: A Comparative Perspective," *Regional Politics and Polity,* 4: 28–66.

Olivesi, C., "The Failure of Regionalist Party Formation in Corsica," in de Winter and Türsan (eds.), *op. cit.*, pp. 174–189.

Reif, K., "Ten Second-Order Elections," in Reif (ed.), *Ten European Elections: Campaigns and Results of the 1979/81 First Direct Elections to the European Parliament.* Aldershot: Gower, 1985, 1–36.

Türsan, H., "Ethnoregionalist Parties as Entrepreneurs," in de Winter and Türsan (eds.), *op. cit.*, pp. 1–16.

Van der Eijk, C. and M. Franklin, *Choosing Europe? The European Electorate and National Politics in the Face of Union.* Ann Arbor: University of Michigan Press, 1996.

CHAPTER 9

The European People's Party and the Restructuring of Right-Wing Parties in Europe

David Hanley and Colette Ysmal

T he European People's Party (EPP) emerged as the clear winner of the 1999 European elections. After four Parliaments, it finally moved ahead of the Socialists and now has a good chance of ending the deal between Christian Democrats and Social Democrats by which they ran the European Parliament (EP) together. Even if the EPP will always need the support of the Party of European Socialists (PES) to steer certain issues in a federalist direction, it is nonetheless in the driver's seat. This outcome is due to several factors, among which the election results of the parties belonging to the EPP in 1994 were not necessarily the most important. What did pay off, in fact, was the EPP's strategy of gaining reinforcements across the board by absorbing parties close to it, but sometimes also parties from further away. This has put the EPP at the heart of the process of rebuilding the moderate right in Europe, but at the cost of losing some of the features that made it distinct.

The EPP: From Christian Democracy to Catchall Party

At the start of the 1994–99 Parliament, the EPP was pursuing a twin strategy. Allied pragmatically with the Socialists in Parliament in order to steer policy in a federalist direction, it was also looking for reinforcements. These were more likely to come, logically speaking, from moderate Conservatives than from Liberals, for whom, as Hugues Portelli remarks, rationalist and secularist traditions still constituted an impediment.[1] But traditional Conservatives were just as suspect to classical Christian Democrats, whether for their excessive enthusiasm for market forces or for their lukewarm attitudes to Europe. Conservatives and Christian Democrats had of course been collaborating since 1978 in a supranational organization, the European Democratic Union (EDU), a branch of the International Democratic Union (IDU), of which many Christian Democratic parties were members while retaining membership in the EPP/CDI (Christian Democrat International). But being close to someone does not imply that one is identical. Thus every new candidate for admission to the EPP, whether it came from the Conservative camp or elsewhere, had to give proof of how it fitted in with EPP ideals. Clearly any judgments about how sincere the candidates were was always going to depend on the balance of forces within the EPP. The German CDU/CSU was reputedly less fussy on such issues than the Italian or Benelux Christian Democratic Parties or the Irish FG (Fine Gael). The breakup of the Italian DC tipped the balance in favor of the Germans, allowing EPP leaders to take a more realistic approach aimed at getting results rather than at doctrinal purity.

Expansion within the European Union

The EPP promptly set out its stall at its December 1994 summit (a conference that brought together EPP heads of government and party leaders), when it decided to grant full membership to "new parties that are not traditional Christian Democrats." By this was meant the Swedish Community and Christian Democratic Party (KDS), the Austrian People's Party (ÖVP), and the Finnish National Rally (KOK), all from countries about to join the European Union.[2] The KDS and ÖVP are in fact described as traditional Christian Democratic parties; they had been associate members while their countries were making the transition to the EU, and it was a mere formality to grant them full membership. It was different for the Finnish KK and the Swedish United Moderate Party (MSP), which also requested full membership at the start of 1995; these are generally considered to be typical right-wing organizations. As they wanted to "remain in the closest possible contact" with the EPP, the latter talked up points of convergence. First, these parties, which had been observers since 1993 (hence not entitled to vote), were totally pro-European and had taken part vigorously in the "yes" campaigns during the referendums held in their countries on joining the EU; second, they agreed with the EPP on numerous points; and last—this was a newer argument, but no less important for that—their electoral strength should not be underestimated: the KOK took 19 percent of the vote, and the MSP 22 percent.[3]

There are two ways of measuring EPP growth: by looking at the Parliamentary group and then at the party itself. In most instances—except for the long-standing case of the British Conservatives, and perhaps the RPR over the next few years—this amounts to the same thing, since a new party tends to join the EPP group and the party per se at the same moment; in some cases there is a gap of a very few years (see table 9.1). In fact the EPP was in the main happy to reinforce its Parliamentary group, thinking in

Table 9.1 National parties' dates of joining EPP

	Full membership	MEPs join group
Austria		
ÖVP (Conservative Christian Democrats)	1995	1995
Belgium		
CVP and PSC (Christian Democrats)	1976	1976
Denmark		
KFP (Christian Democrats)	1995	1992
Finland		
KOK (Conservatives)	1995	1995
France		
CDS/FD (Christian Democrats)	1976	1976
RI/DL (Liberals)		1994
Germany		
CDU/CSU (Christian Democrats)	1976	1976
Greece		
ND (Conservatives)	1981	1981
Ireland		
FG (Christian Democrats)	1976	1976
Italy		
DC (PPI, CDU, CCD, RI) (Christian Democrats)	1976	1976
FI (Liberals)	1999	1998
Luxembourg		
CSV(Christian Democrats)	1976	1976
Netherlands		
CDA (Christian Democrats)	1976	1976
Portugal		
PSD (Liberals)	1996	1996
Spain		
UDC (Christian Democrats)	1986	1986
PNV (Christian Democrats)	1986	1986
PP (Conservatives)	1991	1989
Sweden		
KDS (Christian Democrats)	1995	1995
MSP (Conservatives)	1995	1995
United Kingdom		
Conservatives	–	1992

terms of competition with the Socialists inside the EP. This is an important first step on the way to joining the EPP fully; MEPs from a new party gradually become associated with the work of the EPP and come to seem part of the landscape, as it were, bringing home to their members and voters alike the genuine convergence that they feel exists between their own party and the EPP. Gradually the public comes to believe that the party is already part of the EPP, whereas it is not yet in fact a full member.

Throughout the period 1994–99 the EPP continued to absorb right-wing parties in Europe along three axes, as it had been doing since the 1980s (table 9.1). First, it took in after the 1995 or 1996 elections the four Finnish MEPs of the KOK, the five Swedes from the MSP, and the seven MEPs of the Austrian ÖVP. Second, in November 1996 it recruited the eight MEPs of the Portuguese Social Democratic Party (PSD), who had previously been in the Liberal group. Finally, via the "tradesman's entrance" (joining the Parliamentary group without being let into the party immediately), which had already been used for the British Conservatives in 1992 and the Spanish Popular Party (PP) in 1989, the MEPs of Forza Italia (FI) were accepted, albeit with some difficulty, in June 1998.[4]

By these means the EPP increased its Parliamentary group from 157 members at the start of the Parliament to 201 on the eve of the June 1999 elections.[5] With 32.1 percent of the seats by then, compared with 27.6 percent five years previously, it met the challenge of drawing level with the Socialists, who had only increased their number of seats by 13 thanks to enlargement (from 201 to 214). If this left the Conservatives (by now reduced to their most anti-integrationist elements) in some difficulty, the Liberals found it even harder to make their mark alongside such a dynamic competitor as the EPP, which was just as pro-European as they and increasingly open to market economics.

The Liberals who sat in the EP in the ELDR (European Liberal, Democratic and Reform Party) took their time about setting up a system of affiliated membership, which is a necessary prelude to drawing in established parties that might be hesitant about joining a Europe-wide group. When they did so, they showed a strong propensity for choosing groups with no future (most of which have now disappeared) or were able to pull in only parties with little electoral weight in their own country (the Spanish Social Democratic Center, CDS; the French Radical Party (original *valoisien* version); the Italian Liberal Party and Republican Party). In 1994 they were unable to reach agreement with Silvio Berlusconi and his FI MEPs, who set up their own group, Forza Europea (FE). The Liberals did duly take in the small parties from the new EU member states (the Liberal Forum from Austria, the Center Party (KESK) and Swedish People's Party from Finland,

and the Popular Party from Sweden), but this only brought them 9 extra seats from 1994 to 1999, which is a gain of 1, allowing for the loss of the eight Portuguese MEPs from the PSD.[6] In fact the Liberals went from 15 percent of the vote in 1979 to 6.2 percent in June 1999; at the same time, their number of seats went from 39 out of 410 (9.5 percent) at the first elections held under universal suffrage to a mere 44 (6.7 percent) after the 1999 elections.[7]

Expansion in a Developing Europe

Of course, neither the EPP, nor the Christian Democrat International, nor the CDI's regional organization, the European Union of Christian Democrats (EUCD), confine their activity to the strict framework of the EU. Their networks stretch across the world, and, to take simply the European-level EUCD, they stretch beyond the EU as it was in the 1990s. As we saw, the EPP prepared for the membership of the parties from Austria, Finland, and Sweden that were most likely to join up as soon as these countries became candidates for membership in the EU and, even more, once negotiations were under way. It is just as active with regard to countries likely to be integrated into the EU in future, especially those of central and eastern Europe (CEE) and the Baltic states. The two Hungarian parties—the Popular Democratic Party (KDNP) and the Democratic Forum (RMDF)—which are politically close to the EPP and members of the EUCD and indeed of the EDU, were admitted to the EPP as associate members. It was also made clear that any member of EUCD could obtain similar status once its country was in negotiations with the EU.

A decisive step was taken in July 1996 with the decision of the EPP political bureau at Luxembourg to merge the EPP and EUCD. In October 1996 the EUCD congress laid down seven criteria for harmonizing the two organizations. Some were merely concerned with establishing the ongoing viability of certain parties,[8] but others were important politically or strategically. In terms of doctrine, parties had to give an explicit endorsement of European federalism, personalism (so as to avoid excessive submission to market ideology), and subsidiarity. These principles were, however, to be applied via "a broad definition of Christian Democracy." The aim was to include rather than to exclude. Similarly the 1997 congress accepted that the EPP/EUCD could admit not only representatives from countries actually negotiating entry into the EU but from countries that had simply requested EU membership. The net result of these recruiting operations was impressive. In 1995–96, seven new observers were admitted, and three in 1997; in 1998, three associate members and three observers came into the EPP; in 1999, two associates and two observers.

Some are Christian Democrats, like the two Romanian parties, RMDSZ (the Christian Democratic Party of the Hungarian Minority) and PNCTD (National Farmers and Christian Democratic Party), and the Slovakian KDH (Christian Democratic Movement). As was evident, however, at the Helsinki EPP summit of December 1999, contact was also made with Conservative, Liberal, and peasant parties, according to circumstances. The following groups were thus represented at Helsinki: the Union of Democratic Forces (Conservative) and the Popular Farmers' Union (BANU) from Bulgaria; the Young Democrats' Alliance, or FIDESZ (Liberal), from Hungary; plus the Conservatives of Solidarnosc (AWS) and the Liberals of the Freedom Union (UW), from Poland.

From Identity to Quantity

Casting the net so wide both in the EP and in countries likely to join the EU one day was bound to raise questions about the identity of the Christian Democratic current. By aiming to boost its numbers and become the hegemonic party inside and outside the EU, was not Christian Democracy dissolving itself into a looser grouping that was more representative of the traditional right in either its liberal or conservative version? This debate, which had been ongoing since the early 1980s, was given fresh stimulus when FI joined in 1998. Although the FI MEPs all signed a declaration approving the EPP's values and program, which included the words "Christian Democracy," part of the EPP MEPs were opposed to their joining the group. Broadly speaking, the hard-core supporters of classical Christian Democracy (the Benelux parties, the Irish FG, and the Italian PPI) opposed the choice of these wider parties, first among which was the German CDU/CSU, whose motto might be summarized without too much exaggeration as: "no enemies on the right!" The previous secretary-general of the EPP, the German Thomas Jansen, does not try to sweep such internal disagreements under the carpet.[9] He shows open impatience toward what he regards as the quite unrealistic ideological scruples of some southern European politicians who have still not realized that the future belongs to big, loosely structured center-right groupings that are not too particular about doctrinal matters (apart from paying lip service to the new idol of economic liberalism). Against this, Michael Fogarty deplores what in his eyes amounts to backsliding and sees the EPP as the Trojan horse of neoliberalism.[10]

In order to analyze the current political line of the EPP and its possible differences with the Conservative and Liberal parties, reference will be made to three texts: the EPP manifesto for the 1999 elections; the EDU's Salzburg Declaration, published at its 1998 summit with the upcoming European elections obviously in mind (although this organization for cooperation between Conservatives and Christian Democrats has no official role to play in supporting national parties at European elections); and, finally, the ELDR manifesto, adopted at the Berlin congress of April 1988, *Making the Difference: Unity in Freedom, The Liberal Challenge for Europe.*

It must be said that comparison of these texts shows, with one exception, great similarity between Conservatives and Christian Democrats on basic issues. That said, there are occasional nuances that call to mind the different origins of the parties; these often appear, however, in the form of deliberate omissions, showing a desire to keep quiet about problems that might cause trouble.

On major economic issues, everyone from Conservatives to Liberals agrees about the primacy of the market and the need to reduce the tax burden and encourage enterprise. The state should stick to creating macroeconomic conditions that favor enterprise, particularly by deregulating as much as possible, doing away with the remaining safety nets that the welfare state provides, and providing an adequate labor force via well-targeted education policies. A single currency and an independent central bank are seen as part and parcel of such arrangements.

As for the social consequences of this slide into economic liberalism, there are still vestiges of a "Christian spirit" of compassion toward the most deprived, which is opposed to the logic of, as Pope John Paul might put it, "materialistic, soulless capitalism." The EPP does speak of the need to integrate even the most deprived; the EDU speaks of such people only in a context of deploring the obstacles that government puts in the way of individual success. The EPP is in favor of generous help for refugees, whereas the EDU has nothing to say on this problem. As for social protection, the EPP says clearly that the welfare state is there to help those in need and that redistribution between "rich" and "poor" or between generations should be the norm; the EDU prefers to talk about the negative effect of social charges on jobs and wants to develop private insurance and pensions, abolishing the joint management of social systems (by employers and unions) that exists in some countries.

The tone is the same when discussing the structure of society or social dialogue. Whereas the EDU and to a large extent the Liberals think in terms of a society of individuals free of primary ties (family and nation, for example), free in their work relationships (some people just happen to be employers and others employees), and free to forge links among themselves as individuals, the EPP continues, in the Christian Democratic tradition, to argue in favor of the natural structures of society. These include the family but also corporatist organizations that facilitate dialogue among "social partners" duly represented by approved organizations, such as employers' federations or trade unions, who share a consensual view of society.

If there are undeniable differences of emphasis on economic and social problems between Christian Democrats and Conservatives, the latter are better equipped to set the agenda insofar as they are more coherent, ideologically speaking. The Christian Democrats, wishing as they do to stick to their philosophy as far as possible, have fallen into a semantic trap, still trying to argue in favor of what they used to stand for while in fact capitulating to Liberal and Conservative slogans. What is going to happen, for instance, to the welfare state or "the social market economy," even in watered-down form, when the EPP manifesto proclaims more often than not that "the state makes social problems worse rather than solving them"?

The EPP's unease on social and economic questions is just as great on the future of Europe. There is obviously some consensus on the main tasks awaiting EU foreign policy and on the need for enlargement. The partnership with the U.S. to defend Europe is not questioned, nor is the need for a "European space" in legal and security matters to combat organized crime, illegal immigration, and "terrorism." No one has any difficulty with the idea of a Europe of 20 or 30. But the way Europe functions now and how it might in future is surrounded by deliberate omissions.

In 1998 and 1999 the Liberals and the EPP were explicit in their desire to see the EU develop "coordinated institutions." Nowadays the EPP is less verbose than in its 1992 Athens Declaration and is generally in retreat from its historic positions; it no longer speaks of "federalism," so as to spare the blushes of Conservatives, especially the U.K. Tories, who are in the majority Euroskeptical if not Europhobic. There is no reference to qualified majority voting, and only one to subsidiarity. At the same time, the EPP seems to be feeling pressure from Liberal demands for European institutions to become more democratic and transparent, and in particular for the realization of the principle of a federal government coupled with decentralization that would give countervailing powers to the regions.

Recently, differences between the EPP and Conservatives on the future of European integration have been played down, with attempts being made to handle them as consensually as possible. On social and economic questions, the differences are of degree rather than of kind, and the coming together has been on the terms of the traditional right. The latter might now seem keener on building Europe, but it stands for a model of Europe based on the superiority of competition over social concerns. The fact is that it is becoming harder and harder to maintain the legacy of the "social market" inherited from that "Rhineland capitalism" so cherished by Christian Democracy.

An Election Victory that was Less than it Seemed?

To evaluate the EPP's election results in 1999 and compare them with previous polls, fixed points of reference are necessary. Two such points might be envisaged if we allow for the party's expansion in 1994–99: Either we take the

European election results of 1994 (or 1995–96 for the three countries that joined the EU) for only those parties that joined the EPP directly, or else we make a fictitious interpretation of what those results would have been for parties that came in by other routes. Logically the first solution seems preferable, insofar as parties that came into the EPP from a freestanding position in the EP, such as Forza Italia, or from the Liberal group, such as the Portuguese PSD, could not be considered as part of a family that they opposed at the time.

The French case presents another problem. In 1994 the joint RPR/UDF list led by Dominique Baudis took 25.6 percent of the vote and elected 28 MEPs, who divided into three groups: 13 UDF went into the EPP, 14 RPR went into the Union for Europe of the Nations (Conservative), and the one Radical went into the Liberal group. Such a split obviously cannot be translated into electoral terms, and we are left with choosing the lesser of two evils. Either we do not credit any votes to the two parties—the CDS (nowadays Force Démocrate) and the Parti Républicain (nowadays Démocratie Libérale)—representing the EPP, which is absurd; or we credit all the votes for Baudis's list to the EPP, which is plainly wrong, since the RPR, in spite of the commitment it made in 1994, never joined the EPP group. We choose the second system of reference because it fits in with the 1994 agreement between the UDF and the RPR and because it looks ahead to the 1999 situation.

Having thus established the parameters of our analysis, it seems from table 9.2 that the EPP enjoyed undisputed success across the 15 EU states,

Table 9.2 EPP results in 1994 and 1999

	% vote cast	% vote cast	Diff. 99/94	Seats 1994	Seats 1999	Diff. 99/94
Austria*	29.6	30.7	+1.1	7	7	=
Belgium	24.2	18.7	−5.5	7	5	−2
Denmark	17.7	8.5	−9.2	3	1	−2
Finland*	20.2	25.3	+5.1	4	4	=
France	25.6	22.1	−3.5	13	21**	+8
				(27)		(−5)
Germany	39	48.7	+9.7	47	53	+6
Greece	32.7	36	+3.3	9	9	=
Ireland	24.3	24.6	+0.3	4	4	−
Italy	10	37	+27	12	37	+25
Luxembourg	31.5	31.7	+0.2	2	2	=
Netherlands	30.8	26.9	−3.9	10	9	−1
Portugal	−	32.2	+32.2	1	9	+8
United Kingdom	26.8	35.8	+9	19	36	+17
TOTAL	28.7	37.6	+9.9	173	233	+60

*1995 for Sweden; 1996 for Austria and Finland.
**Includes votes for the list of Messrs. Sarkozy and Madelin (RPR-DL).

since it went from 28.7 percent of the vote in 1994 and 1995–96 to 37.6 percent in 1999, and from 173 to 233 seats. It might be thought that this success was something of an illusion or, to put it differently, that it owed a lot to the changing shape of the EPP between 1994 and 1999 as well as to the RPR's input in 1999. But this is not so. The parties that belonged to the EPP in 1994 were already on a roll.

The Electoral Consequences of Expanding the Group

This is obviously crucial, because the Portuguese PSD, which was not present in 1994, took 32.2 percent of the vote in 1999, and Forza Italia took 25.2 percent. In terms of seats, there was also significant progress, since these two recruits brought the EPP some 30 seats. But these were part of the EPP total before the 1999 elections (see above). Also, this balance sheet is much less favorable if we compare the results of these two parties as well as those of the RPR/UDF tandem in France in 1994 and 1999. These three parties, taken together, won 29.1 percent of the votes cast in 1994; in 1999, only 25.8 percent. The relapse was general, as Forza Italia went down from 30 percent of the vote in 1994 to 25.2 percent five years later; the PSD slipped from 35.6 percent to 32.2 percent, while the RPR/UDF, united in 1994 but divided in 1999, went from 25.6 percent to 22.1 percent. In terms of seats, even if the PSD remained stable with 8 seats, FI lost 5 (from 27 to 22), while even if the RPR's entry enabled the EPP to raise its French contingent from 13 seats to 21, this was still a considerable shortfall compared with the notional 1994 tally of 27.

In the end, the hunt for bodies paid off. Yet in terms of electoral influence, the EPP's success was not that great. But instead of blaming it for that, it is better to look back at the different national contexts, which in European elections are always much more significant than European issues and hence the European labels that parties wear in the European Parliament. Forza Italia's relative failure goes back to the victory of the Olive Tree center-left coalition in the 1996 parliamentary elections in Italy, even if Silvio Berlusconi's party did improve on its score in that contest (20.6 percent). Similarly, in Portugal, the PSD has been struggling since its 1995 general election defeat, when the Socialist Party took over; we may note that these Conservatives were down 2 percent on their 1995 score (34.1 percent). Finally, in France, the slippage of the UDF and RPR is connected with both the crisis that these parties have been experiencing since the failed dissolution of the National Assembly in 1997 and the emergence of a "sovereignist" current of opinion represented by Charles Pasqua and Philippe de Villiers, which took 13.1 percent of the votes cast.

The Old EPP Parties: Contrasting Results?

Referring now to the election results of parties that were in the EPP in 1994, we find that the position of the Christian Democrats and Conservatives is much more favorable in general. In 1994, 1995, and 1996 they won 29.2 percent of the vote; in 1999, they won 31.4 percent, a gain of 2 percent. This resulted from a combination of stability in some countries, losses in others, and gains in a limited number of states, which however counted heavily enough in terms of their electoral weight, numbers of seats, and symbolic images to make the EPP the winner of the European elections.

As table 9.2 makes clear, there was very little movement upward or downward (between −0.2 percent and +1.3 percent) from 1994 to 1999 for Austria, Spain, Ireland, Luxembourg, and Sweden. Italy can be added to the list if we separate the results of the Christian Democrats, represented in 1999 by five lists [the Italian People's Party (PPI), United Christian Democrats (CDU), Christian Democratic Center (CCD), Union of European Democrats (UD.EUR), and, finally, Dini's Italian Renewal (RI)], from those of Forza Italia. The five lists together won 11.8 percent of the vote in 1999, whereas the PPI alone had taken 10 percent five years before.

In other countries, however, the pattern is much more varied. There was decline in the Netherlands (almost 4 percent), Belgium (almost 6 percent), and Denmark (9 percent). But there were some particularly important gains: over 3 percent in Greece, over 5 percent in Finland, and above all over 9 percent in Germany and nearly 10 percent in the U.K.

These successes and failures depended above all on national contexts.

We shall take four examples. The first is Denmark, where the slippage of the Christian People's Party (KRF) seems to be connected to the advance of anti-European parties like Rally for Justice and of the populist far right, represented by the Danish People's Party (DF), which took 5.8 percent of the votes cast at its first electoral outing. The KRF, which is very favorable to European integration but very conservative in economic, social, and cultural matters, faced competition for its voters, in a country that is more and more wary of the EU, from parties that borrowed its nationalist themes and related them to a series of problems (taxes, bureaucracy, immigration) for which they laid the blame on European integration.

The second example is Belgium, where the result of the 1999 European elections was bound to be linked to national issues, as the European poll took place on the same day as elections for both houses of parliament and the regional councils.[11] The election campaign swept the European dimension completely under the carpet. Far from giving a verdict on the future of European integration, the powers of the Commission, or the EP, Belgian voters were voting about the various crises affecting the kingdom, from the crisis of federalism, which has been obvious since the early 1980s through the dioxin scandal, to the Dutroux case, which exposed the malfunctioning of the police and judicial system. The Parliamentary and European votes were a punishment for the parties that had been in office (Socialists and Christian Democrats), but more for the latter than the former. In the Dutch-speaking electoral college, the Christian People's Party (CVP) went down five points from 27.4 percent in 1994 to 21.7 percent in 1999, and Jean-Luc Dehaene's party even finished behind the Liberals. There was a similar drop in the French-speaking college, since the Christian Social Party (PSC) fell from 18.8 percent in 1994 to 13.3 percent and lost its place as the third party in Wallonia to the ecologists. In Belgium as a whole, the Christian Democratic Parties are no longer the biggest political force in the country and—though this is not strictly speaking a consequence of the European elections—now find themselves in opposition after 40 years of power sharing, usually with the Socialists.

Anyone could have foreseen the success of the CDU/CSU in Germany. Since winning the general election of 1998, which made Gerhard Schroeder chancellor, the SPD had been showing internal divisions and picking up a string of defeats in regional or local elections. The signing of the manifesto "For a Social and Liberal Europe" shortly before the elections by the German chancellor and the British prime minister Tony Blair, which suggested a "third way" for socialism, served only to increase tensions in the party and in the Parliamentary group, as well as sharpening the disappointment of voters whose expectations in 1998 had been only very partially met by a government whose economic and social policies they did not find very helpful.

Although it was expected, the scale of the vote against the coalition in office (SPD and ecologists) reached proportions beyond what the Christian Democrats had hoped for. With 48.7 percent of the vote and 53 seats, the CDU/CSU had their second-best result since the first election of the EP by universal suffrage in 1979 (49.2 percent). They gained ten points from 1994 (39 percent), and that had been a vote that already showed how exhausted the government was after 12 years in office. As Stephanie Abrial has shown, the Christian Democrats did not just improve their overall score but made gains in all the *länder*.[12] Thus, in comparison with 1994, they strengthened their position in the regions where they traditionally do best (Baden-Würtemberg, from 42 to 50.9 percent; Bavaria, from 48.9 to 64 percent; Saxony, from 39.2 to 45.9 percent; Schleswig-Holstein, from 40.6 to 50.5 percent), but they also won votes in the most leftward-inclined *länder*. In Saarland they went from 35.6 to 44.9 percent, to finish ahead of the SDP; they gained over 10 percent (from 37 to 47.3 percent) in Rhineland-North Westphalia. They also made significant gains in Hamburg, going from

32.1 percent of the votes cast to 40.2 percent. Overall the CDU/CSU beat the ruling coalition of SPD and Greens in 11 of the 16 German regions.

If the European elections in Germany fulfilled their role as "midterm elections," encouraging the expression of voters' displeasure and disappointment and strengthening opposition parties, the British situation seems different, in spite of the Conservatives' victory. With 34 percent of the vote, the Tories effectively canceled out their very poor 1994 score (27.9 percent) and gained slightly more than three points on their 1997 general election score (30.7 percent). They became once again the number one party in the U.K., albeit on a hugely reduced turnout. As Céline Belot and Bruno Cautrès note,[13] the Conservative Party came in first in 6 regions out of 11, particularly in the south of England, which was the basis of Tory success in the 1980s.[14] This result is all the more remarkable as the Conservatives were in competition with the U.K. Independence Party, which favors British withdrawal from Europe and took 6.7 percent of the vote.

The Conservative revival might seem surprising if we refer to the high levels of popularity enjoyed by Tony Blair and the persistent disaffection with the Conservatives. In a MORI poll of late May 1999, 58 percent of those questioned declared themselves satisfied with the prime minister, and 52 percent declared themselves ready to vote Labour in a general election. By contrast, only 20 percent declared themselves satisfied with the Conservative leader William Hague, and only 28 percent intended to vote Tory. Yet the European elections seemed likely to prove more difficult for New Labour, as an ICM poll of June 8 for the *Guardian* put the gap between the two parties much lower. Two days before polling day, 31 percent of those who were certain that they would vote chose the Conservatives, and 38 percent Labour.

These hints, which were belied by the actual results, seem to show up the real meaning of the European elections in Britain. Unlike the case in Germany, it was not a test of Tony Blair's domestic policy, but rather a sort of referendum on British options over Europe. While the prime minister had shown a more open attitude to Europe, albeit prudent and very pragmatic, the Conservatives had in contrast stressed the Euroskeptical line taken by William Hague. Facing a Blair who seemed to be envisaging the adoption of the single currency some time in the period 2003–2005, following a referendum, the Conservative leader centered his campaign on the eventual abandoning of the pound and warned that the European elections would be the real referendum on this question. The Conservative success was therefore built on the lack of enthusiasm felt by the British toward European integration, on their attachment to their national currency, and thus on their long-standing and still flourishing Euroskepticism.

Taking as a baseline the EPP's total of seats at the end of the 1994–99 Parliament (201), it gained a definite 32 seats, including 18 in the U.K, and

6 in Germany. This was an important and symbolic victory, but it is by no means sure that it will really increase the group's cohesion, especially if one adds in the presence of RPR MEPs who are more European than they used to be but are still wary of federalism. But the federal idea seems to have lost ground in the German CDU also.

The 1999 results also increased the weight of Conservatives and Liberals over the original Christian Democrats, especially the Dutch, Belgians, Italians, Luxembourgers, and Irish, who are most attached to CD values. The way in which the EPP is developing, set by the CDU/CSU, which took the lead in moving closer to the other parts of the moderate right, seems destined in the long run to do away with the originality of Christian Democracy or to dissolve it into a vast conservative "camp," much more liberal and less socially inclined than Christian Democracy ever was.

This rapprochement or convergence might have some effect on supranational bodies. At the end of 1999, 24 of the EPP's 36 members and associates were simultaneously members of the EDU, which has only 43 members, including a number of small parties that do not amount to very much. Two EDU heavyweights—the Tories and the RPR—are members of the EPP group. The fact that the group is now called EPP-ED (European Democrats) shows how porous the frontiers are between the EDU and the EPP. Within the European Parliament there are now few forces on the right that manage to exist outside the EPP; this necessarily raises questions about the two Internationals and their regional structures. A merger of the two transnational bodies no longer seems totally out of question; the new body would have a Christian Democratic label, but it would cover a Conservative reality. This might be one of the hidden consequences, albeit not immediate, of the 1999 European elections.

Notes

1. Portelli, H. "La récomposition des droites modérées," in Perrineau, P., and C. Ysmal (eds.), *Le vote des Douze. Les élections européennes de juin 1994*, FNSP/Département d'Etudes Politiques du Figaro, 1995.
2. Party names and initials are taken from Hermet, G., J.-T. Hottinger, and D.-L. Seiler, *Les Partis politiques en Europe de l'Ouest*, Economica, 1998.
3. The EPP was also in contact with the Norwegian conservative party Höyre (the Right), which was originally considered in the same light as its opposite numbers in Austria and Scandinavia but could obviously not be fully integrated into the EPP setup once Norway had refused to join the EU.
4. The bargaining went on in fact for a long time. As early as 1995 the FI members had joined the EPP groups in the Council of Europe, the Western European Union, and the Committee of the Regions. It was necessary to wait for FI to

proclaim itself a democratic party, and not a movement, and for it to distance itself from the National Alliance (AN), which in the eyes of the EPP had "a past with Fascist associations." Even then, joining the group was not a smooth process; each FI MEP was voted on separately by the EPP group, with votes in favor usually between 90 and 96, and votes against in the 34 to 38 range.

5. Johnson, K.-M., *Transnational Party Alliances: Analysing the Hardwon Alliance between Conservatives and Christian Democrats in the European Parliament*, Lund University Press, 1997.

6. We proceeded as if the MEPs of the PR (nowadays Démocratie Libérale) elected for the RPR/UDF united list in 1994 had registered immediately with the EPP. They committed themselves to doing so during the election campaign, but it was December 1994 before they were "invited to withdraw from ELDR," essentially for having moved too close to the EPP.

7. *Le Monde*, June 15, 1999.

8. They were asked to have won at least 10 percent of the votes cast in the last elections to their national parliament or at least 5 percent over the last two such elections; to have had members sitting in these parliaments; and not to have been affected by splits for two years.

9. Jansen, T., *The European People's Party: Origins and Development*, Macmillan, 1998.

10. Fogarty, M., *Motorways Merge: The New Challenge to Christian Democracy*, Christian Democrat Press, 1999.

11. Pina, C., "Belgique: crise politique et victoire par défaut des libéraux," *Revue politique et parlementaire* 1001, July–August 1999.

12. Abrial, S., "Allemagne: une sanction lourde pour Gerhard Schröder," *Revue politique et parlementaire* 1001, July–August 1999.

13. Belot, C., and B. Cautrès, "Royaume-Uni: une élection sans électeurs, mais non sans conséquences," *Revue politique et parlementaire* 1001, July–August 1999.

14. Note that the electoral system changed between 1994 and 1999. The single-ballot first-past-the-post system in 84 constituencies in the U.K. was replaced by a proportional system based on 12 regions. Only Northern Ireland kept the same territorial arrangements and the same mode of voting, as it had already elected its MEPs by PR in 1994. These changes rule out any comparisons in terms of seats.

CHAPTER 10

The Extreme Right in the June 1999 European Elections

Piero Ignazi and Pascal Perrineau

An analysis of the European party system shows that extreme right-wing political parties were basically nonexistent until the beginning of the 1980s. Looking at the results of the first European elections in 1979 (see table 10.1), only one extreme right-wing party entered the European Parliament: the Movimento Sociale Italiano (MSI). Other parties participated in those elections: the Parti des Forces Nouvelles (PFN) in France, the Nationaldemokratische Partei Deutschlands (NPD) in West Germany, the National Front (NF) in Great Britain, the Vlaams Blok (VB) in Belgium, and the Fremskridtsparti (Progress Party, FRP) in Denmark. Whether this Danish party, along with its Norwegian counterpart, should be included in the extreme right, however, remains debatable; and the Democratic Unionist Party (DUP), created in 1971 by the Presbyterian pastor Ian Paisley, is more the result of an ultraconservative and intransigent Protestant movement than of the extreme right.

This situation changed profoundly in the 1984 European elections. Three extreme right-wing parties won representation in the European Parliament in Strasbourg. The MSI was joined by the French National Front, which won more votes (11.2 percent compared to 6.5 percent) and more deputies (ten compared to five) than the Italian party. The small Greek party EPEN also entered the Parliament with 2.6 percent of the vote and one seat. Furthermore, even though it did not win any seats, the Centrumdemocraten Party (CD) in the Netherlands obtained its best score with 2.5 percent of the vote. These elections changed the extreme right-wing political landscape. New political parties appeared, while other established parties became stronger following a period of insignificance and obscurity. As a result of the National

Table 10.1 The evolution of the extreme right in the European Parliament, 1979–1999

	1979			1984			1989			1994			1999		
	Party	Seats	Group	Party	Seats	Group	Party	Seats	Group	Party	Seats	Group	Party	Seats	Group
Germany							REP[3]	6	GTDE						
Austria										FPÖ	6	NI[5]	FPÖ	5	NI
Belgium							VB	1	GTDE	VB	2	NI	VB	2	NI
										FN	1	NI			
Denmark	FRP	1	DEP										DFP	1	UEN
Spain															
Finland															
France				FN	10	GDE	FN	10	GTDE	FN[4]	11	NI	FN	5	NI
Greece	KP	1	NI[1]	EPEN	1	GDE									
Ireland															
Italy	MSI-DN	4	NI	MSI-DN	5	GDE	MSI-DN	4	NI	MSI-AN	11	NI	MSFT	1	NI
Luxembourg															
Netherlands															
Portugal															
Great Britain	DUP	1	NI	DUP	1	NI	DUP	1	NI	DUP	1	NI	DUP	1	NI
Sweden															
Total		7			17[2]			22			32			15	

[1] 1981.
[2] In 1987, the DUP deputy (Northern Ireland) joined the GDE.
[3] As of 1990: 3 REP (ND); 3 DL (GTDE).
[4] At the end of the term: 10 FN (including two shift-overs); 2 MN; 1 miscellaneous extreme right.
[5] 1996.

Front's success in France, the extreme right was able to create the European right-wing group in the European Parliament. The 16 deputies of the FN, MSI, and EPEN were joined by one Unionist deputy from Northern Ireland.

This modification of the political landscape would be confirmed at the end of the 1980s and beginning of the 1990s. Numerous signs testified to this change—the sudden success in 1989 of Republikaner (7.6 percent of the vote and six deputies); the comeback at the local level of the NPD and the Deutsche Volksunion (DVU) in Germany; the establishment of the VB party (4.5 percent of the vote and one deputy) and the resurgence of the French-speaking extreme right in Belgium; the survival of the CD in the Netherlands; and finally the confirmation of the Danish FRP's strength (5.3 percent of the vote) before its internal turbulence. These gains, associated with the establishment of the FN in France (11.7 percent of the vote and ten deputies) and the MSI's persistence in Italy (5.5 percent of the vote and four deputies), demonstrated that an extreme right-wing dynamic was at work in Europe. A group bringing together the FN, the VB, and the Republikaner, but excluding the MSI—considered too "Europeanist" and too opposed to German nationalists on the issue of autonomy for Alto Adige—was reformed under the direction of Jean-Marie Le Pen. In Europe, one can also point to the proliferation of extremist parties in Switzerland (Swiss Democrats, Automobile Drivers Party, Ticino League); the growth of the Norwegian Progress Party; and the triumphal march of the Freiheitliche Partei Österreich (FPÖ) in Austria, on its way to becoming the strongest extreme right-wing party in the 1990s.

This trend showed its first signs of weakness in the 1994 European elections (see table 10.2). The FN's support eroded slightly in France (10.5 percent of the vote and eleven deputies) and the Republikaner Party in Germany no longer attained 5 percent of the vote (3.9 percent). The extreme right registered gains only in Belgium (7.8 percent of the vote for the VB with two deputies, and 2.9 percent of the vote for the Belgian FN with one deputy). The MSI, which became the MSI- Alleanza Nazionale (12.5 percent of the vote with eleven deputies), slowly left the extreme-right sphere. Because the MSI refused to join an extreme right-wing Parliamentary group, Jean-Marie Le Pen could not form a group in the European Parliament, and the extreme right wound up in the same group as the independents. In 1996, following Austria's admission to the European Union, Jorg Haider's FPÖ won 27.5 percent of the vote and elected six deputies, but, given the concern for "respectability," it was out of the question here as well to allow the national-populist Austrian to join a Parliamentary group with the FN and the VB.

Many attempts to define and classify the extreme-right in studies were made in the 1990s. One classification distinguished between the parties of

Table 10.2 The evolution of the extreme right in the European elections, 1994–1999

Country	Party	% 1994	Deputies	Group	% 1999	Variation	Deputies	Variation	Group
Germany	Die Republikaner (Rep)	3,9	–		1,70	-2,20	–		
	Nationaldemokratische Partei Deutschlands (NPD)	0,2	–		0,40	0,20	–		
Austria	Freiheitliche Partei Österreichs (FPÖ)/ Die Freiheitlichen	27,53 (1996)	6	NI	23,48	-4,05	5	-1	NI
Belgium	Vlaams Blok (VB)	7,8	2	NI	9,40	1,60	2	=	NI
	Front National (FN)	2,9	1	NI	1,50	-1,40	–	-1	
	AGIR	0,7	–						
	Belgique-Europa-België (BEB)	0,4	–						
	Front nouveau de Belgique (FNB)	–	–		0,40		–		
	Parti Communautaire National-Européen (PCNE)	–	–		0,00		–		
Denmark	Fremskridtpartiet (FRP)	2,9	–		0,70	-2,20	–		
	Dansk Folkeparti (DFP)	–	–		5,80		1	+1	UEN
Spain	Falange Española de las JONS	0,06	–		0,07	0,04	–		
	Falange Española Independiente (FEI)	0,03	–						
	Comunion Tradicionalista Carlista (CTC)	0,03	–						
	Democracia Nacional (DN)	0,03	–		0,04	0,01	–		
	Alianza por la Unidad Nacional (AUN)	–	–		0,06		–		
	La Falange (FE)	–	–		0,05		–		

Country	Party						
Finland	Vapaan Suomi Liitto (VSL)	0,6 (1996)					
France	Front National (FN)	10,51	11	5,69	−4,82	5	−6
	Mouvement National (MN)	–	–	3,28		–	
	Ligue nationaliste (LN)	–		0,01			–
Greece	Ethniki Politiki Enossis (EPEn)	0,8	–	0,76	+0,66	–	–
	Proti Grammi	0,1		0,26		–	
	Komma Ellinismou	–		0,12			–
	Eliniko Metopo	–				–	
Ireland		NI		NI			
Italy	Movimento Sociale Fiamma Tricolore (MSFT)	(AN)	1	1,60		1	=
Luxembourg	National Bewegong	2,4					
Netherlands	Centrumdemocraten (CD)	1,1	–	0,50	−0,60	–	
Portugal		?		0,96			
Great Britain	British National Party (BNP)						
Sweden	Ny Demokrati (ND)	0,1 (1995)		0,34		–	
	Sverigedemokraterna (SD)						
Total		21		14			−7

the old, traditional extreme right and a new postindustrial extreme right.[1] This last category emerged from the observation of numerous extreme-right parties that met with electoral success in the 1980s, of which the French FN is the prototype, and introduced values and ideological characteristics on the political scene different from those traditionally supported by neofascist parties. This typology also emphasized the electoral decline of neofascist parties (and in some cases even their disappearance) on one hand, and the consolidation and development of new parties on the other. The accentuation of this dual trend over the past several years has made the distinction between traditional and postindustrial parties nearly obsolete. By the end of the 1990s the traditional, neofascist extreme right had for all practical purposes disappeared. In those countries where authoritarian extreme right-wing regimes had survived until the 1970s (Spain, Portugal, Greece), the hereditary parties of the defunct regimes suffered a veritable fiasco in the first democratic elections. In Spain, the tranquil and negotiated transition toward democracy (marked by the culture of pacts and compromise) prevented the birth of a party nostalgic for the Franco regime.[2] The only party that entered the political scene in an effort to revive Franco's heritage was Fuerza Nueva. Led by Blas Pinar, a marginal Francoist figure, this party without any real structure or political program other than pure and simple nostalgia for the Franco regime never succeeded in achieving anything but electoral insignificance. It participated in the first democratic elections (1977) associated with other extreme right-wing forces under the banner 18th of July Alliance (the day the insurrection against the Spanish Republic was launched) and obtained a ridiculous 0.35 percent of the vote. It was not until the next elections, in 1979, that Fuerza Nueva (this time under the label National Union) succeeded in electing one deputy (Blas Pinar) to the Cortes and reached 2.1 percent of the vote. The weakness of Fuerza Nueva, confirmed by its electoral failure in the 1982 elections, led to its dissolution. This failure was also due to the fact that another political actor, much more solid and established, the Alianza Popular (AP), founded by Fraga Iribarne, a former Franco minister, occupied the political space on the right (including the extreme right). During those years, the positions adopted by the AP, at times quite radical, left doubts as to whether that party should be classified as conservative. Already challenged on the extreme right by the AP, Fuerza Nueva and the parties that succeeded it (such as the National Front, founded by Blas Pinar in 1986) were also forced to confront populist candidates and political formations such as those of the businessman Ruiz Mateos (who attracted 3.9 percent of the vote in the 1989 European elections) and later of the controversial mayor of Marbella Jesus Gil y Gil. In the end, the Spanish extreme right did not manage either to keep supporters of the

Franco regime alive or to create a durable extreme right based on the Le Pen model. Following the attempt to impose a National Front, the last effort in this direction was the creation of National Democracy in 1995. This party tried to project a more modern and moderate image, paying particular attention to social problems. This last effort also failed, and in the 1996 legislative elections, the Spanish extreme right, with 0.05 percent of the vote, registered one of the weakest scores in Europe.

The Greek case is characterized by a similar situation. After the fall of the "colonel's regime" (1967–1974), different political formations inspired by neofascism tried their hand, proclaiming the positive role played by an authoritarian regime. From the first democratic elections in November 1974, the Democratic National Union (EDE) presented itself in this political space, obtaining 1.1 percent of the vote but failing to win a single seat in the legislature. In the 1981 European elections, the Progress Party (KP), founded in 1979, had a more enviable destiny, acquiring 2 percent of the vote, which allowed it to send one deputy to Strasbourg. Nevertheless, in the legislative elections that took place at the same time, the party attained just 1.4 percent of the vote, and not a single deputy from the party was able to sit in the Parliament. A new party, the National Political Union (EPEN), won 2.3 percent of the vote and one deputy in the 1984 European elections. But following this modest success, the Greek extreme right declined. In the 1989 European elections, the EPEN, with 1.2 percent of the vote, lost its only seat in the European Parliament and won less than 1 percent of the vote in the elections of 1993, 1994, and 1996. The extreme right wing's efforts to exploit the nationalist wave during the last few years, particularly with regard to anti-Macedonian sentiment, have not allowed it to emerge from the electoral fringe. The "nostalgia" of the EPEN, built on defending the colonels' regime, the absence of modern political entrepreneurs, and the unifying character of the traditional right-wing party, Nea Dimokratia, which occupies all the space on the right, have all been inhibiting factors for the extreme right.

The Portuguese case is slightly different from the two preceding ones. As in Spain and Greece, the partisans of the Salazar regime gave birth to a party of nostalgia during the transition to democracy under the impulse of a veteran of the colonial wars, General Kaulza de Arriaga. Baptized the Independent Movement for National Reconstruction (MIRN), it was later renamed the Party of the Portuguese Right (PDP). Despite a certain moderation in tone and an alliance with the Christian Democratic Party (PDC), the PDP never achieved significant electoral results, and the MIRN disintegrated. Only the PDC remained to try and occupy the place of the extreme right. This traditionalistic Catholic party, hostile to the principles of liberty and equality, was never able to attract even 1 percent of the vote (0.7 percent in the 1987 and 1989 European elections), and it appeared to have come to its end when it was not able to present a list of candidates in the 1991 elections. No other movement—neither the Forza National-New Monarchy Party (FN-NM), a pale imitation of the Le Pen movement, nor the radical Movement of National Action (MAN)—has managed to acquire any political weight whatsoever. In conclusion, the countries that have recently experienced authoritarian regimes that included some characteristics of fascist states have seen only the stillbirth of parties claiming inspiration from that tradition and have not succeeded in establishing a new type of extreme right-wing party.

In the Italian case, the traditional extreme right modeled on neofascism has been totally eclipsed. Immediately following World War II, the MSI, founded in 1946 and present in the Italian Parliament from the first elections in 1948, imposed itself among the extreme right in Europe. Since then it has been continually represented in parliament, gathering on average between 5 to 6 percent of the vote (at times becoming the fourth largest Italian party) and developing a well-structured organization made up of hundreds of thousands of members and thousands of elected officials. Until the early 1980s, all of these characteristics put the party in a totally different situation from the mostly tiny extreme right-wing parties, with the partial exception of the German NPD during the second half of the 1960s. But, at the end of the 1980s, when the National Front was asserting itself in France, the MSI began a slow decline. In the 1990 regional elections, the party received less than 4 percent of the vote for the first time since 1948. To face the debacle, the party's leadership was reorganized. Gianfranco Fini, the young chosen successor of the historic leader Giorgio Almirante, was called back to head the party in July 1991. At first, a return to neofascist orthodoxy was advocated. Then, in 1994, the MSI suddenly set out on a different road from neofascism. In the fall of 1993, at a time when the entire party system fell apart and the Italian political class was discredited, MSI candidates registered real electoral successes in Rome and Naples and for the first time received the most votes of any party on the right. Added to this unexpected electoral breakthrough was the introduction of a majority electoral system for three-fourths of electoral seats in the upcoming legislative elections of March 1994. In this new context, the party decided to abandon the old MSI acronym to become the Alleanza Nazionale (AN), while at the same time softening nostalgic references to the old Fascist regime. The electoral success of 1994—13.5 percent of the vote—propelled it to become the country's third most important political force, allowing it to enter Silvio Berlusconi's right-wing government and causing its transformation to accelerate. At the Fiuggi Congress in January 1995, the party officially changed its name to Alleanza

Nazionale and repudiated fascism despite some residual ambiguities. In keeping with this new ideological course, the party refused all relationship with the National Front and remained isolated in the European Parliament in Strasbourg. This evolution, confirmed and amplified by the political theory conference in Verona in February 1998, put the AN out of the orbit not only of the traditional extreme right wing, but also of the extreme right as a whole. The AN is not a racist, anti-immigrant party, it is not hostile to capitalism and free enterprise, it is not antidemocratic or Euroskeptical. It became, at least at the level of its leadership, a member of the conservative and moderate political family. The flame of the traditional extreme right is maintained in Italy today by the MSI-Fiamma Tricolore, created in 1995 by MSI members who rejected their party's transformation into the National Alliance. This traditional fascist party, headed by former MSI leader Pino Rauti, obtained 1.7 percent of the vote in the April 1996 legislative elections. The party, which defines itself as a force of "national-populism in its content and national-revolution in its strategy," employs only marginally the xeno- phobic themes largely utilized by other extreme right-wing movements. It privileges social themes and vigorously opposes economic liberalism and globalization. Its supporters come mainly from poor urban suburbs and the traditional bastions of the south. Nevertheless, internecine quarrels seriously limit the party's potentialities. Even the drop in score registered by AN during the last European elections (10.3 percent of the vote compared with 12.5 percent in 1994) did not leave a real electoral space for MSI-FT, which attracted only 1.6 percent of the voters. This marginalization of the tradi- tional extreme right even in Italy, which had been its chosen country, is confirmed in Europe today when one considers the British case.

In Great Britain the extreme right has never found a place in the political spectrum. The reasons for this are at once cultural (the force of constitutional traditions), systemic and institutional (the one-round electoral system and the tradition of alternating power between Conservatives and Labour), and, finally, political and organizational (lack of leadership, internal quarrels, and factionalism within the extreme right itself). Yet beyond these handicaps, the extreme right in Britain has never been able to extract itself from a fascist tendency with a definite aftertaste of neo-Nazism, which has kept it in a political ghetto. Such was the case of the National Front in the 1970s and its successor, the British National Party (BNP), in the 1980s and 1990s. The only time when the extreme right tried to come out of the woods was during the 1979 legislative elections, when the NF presented candidates in numerous constituencies of the kingdom. In the face of this challenge, the Conservative Party presented a series of right-wing proposals in an effort to cut the grass under the NF's feet. Following its failure in the elections, however, the NF descended into marginality. Its obscure internal conflicts and the creation in 1980 of the BNP, a party loyal to neo-Nazism, racial hatred, and revisionist history, did not inverse the trend. From time to time BNP candidates obtained a significant percentage of the vote that went as high as 10 percent or more in local elections. In the June 1999 European elections, the BNP obtained its best results in West Bromwich (4.5 percent of the vote) and in two districts of the greater London area (Dagenham, with 4.7 percent, and Barking, with 4.9 percent). Nevertheless, the BNP does not have any of the characteristics of a nationally established party capable of instituting itself within the British party system. Caught in its references to fascism, Nazism, and other extreme right wing tendencies embodied by the philosopher Julius Evola, the traditional British extreme right has no serious chance of developing.

While fascism does not appear to have heirs of any weight in the European extreme right, the situation is different for the postindustrial extreme right. The latter includes parties such as the National Front (FN) and the National Republican Movement (MNR) in France, the FPÖ in Austria, the Dansk Folkeparti in Denmark, the Republikaners and the DVU in Germany (the third largest extreme right wing party in Germany, the NPD, is closer to the neofascist tradition), the Centrumdemocraten in the Netherlands, and the Vlaams Blok and the FNB in Belgium (despite the fact that these two last parties also have characteristics that liken them to the neofascist family). All of these parties share certain common sociological traits as well as common values. What impact did these parties and their partners have during the last European elections in June 1999?

In June 1999, the extreme right's erosion at the polls was significant (table 10.2). After a decade of growth, the extreme right is experiencing a decline. With the exception of Flemish-speaking Belgium and Denmark, the extreme right and national populism are weakening. In Belgium the VB, with 9.4 percent of the vote, increased its share of the vote by 1.6 points compared with 1994, but nevertheless has not been able to win a third seat in Strasbourg. It reached 15.1 percent of the vote in the Dutch-speaking electoral college, almost one point ahead of the Flemish Socialist Party. In the French-speaking electoral college, the FNB, with 4.1 percent of the vote (1.5 percent in all of Belgium), lost its one seat, while the dissident New Front of Belgium (FNB) gathered only a slim 1.1 percent of the vote (0.4 percent in all of Belgium). The Parti Communautaire National-Européen (PCN-E) had a very timid following (0.1 percent of the vote in French-speaking Belgium and 0.04 percent in the country as a whole). In Denmark, while support for the FRP eroded significantly (2.9 percent of the vote in 1994, just 0.7 percent in 1999), the Dansk Folkeparti (DFP), created in 1995 by Pia

Kjaersgaard following her eviction from the FRP leadership, confirmed its success in the 1998 legislative elections (7.4 percent of the vote and 13 seats) by getting 5.8 percent of the vote in 1999. The DFP's best scores were in the Roskilde Amtskreds (7.3 percent of the vote) and Kobenhavns Amtskreds (7.2 percent) districts. The DFP elected one deputy, who sits in Strasbourg as a member of the "sovereigntist" group, Union for Europe of the Nations. In 1979, the one deputy elected by the FRP had joined a similar group rather than joining his extreme right wing colleagues who at the same time joined the independent group. This is a characteristic of the extreme right wing parties of northern Europe. While sharing a xenophobic platform with their other European partners, they have their base in a public and antifiscal protest movement hostile to the Scandinavian welfare states.

A populist formation of this type had some electoral successes in Sweden at the beginning of the 1990s. Ny Demokrati (New Democracy), founded nine months before the legislative elections of 1991, obtained 6.7 percent of the vote and elected 25 deputies to the legislature. Little by little, however, a xenophobic current imposed itself that brought about the ND's decline. The party gathered just 1.2 percent of the vote in the September 1994 legislative elections and 0.1 percent in the European elections of September 1995. In 1999, Sverigedemokraterna (the Swedish Democrats (SD) as of 1988), heir to the xenophobic movements Bevara Sverige Svenskt (Keep Sweden Swedish, 1979) and Sverige Partiet (Party of Sweden, 1986), was unable to emerge politically (0.33 percent of the vote) and at the same time was subjected to competition from other activist and xenophobic movements, the radical dissidents of Hembygdspartiet (Party of the Country, 1995). The same weakness of the extreme right is to be found in the Netherlands, where Hans Janmaat's party, the Centrumdemocraten (CD), gathered only 0.5 percent of the vote and lost 0.6 point since 1994.

Farther south, in the three countries that had experienced authoritarian extreme right wing regimes until the early 1970s, nostalgia for the defunct governments found an electoral echo only among a minuscule minority of voters. The four extreme right wing parties in Spain gathered just 0.19 percent of the vote. The various Greek extreme right wing lists collected about 1 percent, while the two extreme right wing Portuguese parties (which in fact belonged more to the traditional right than to the extreme right) gathered less than 1 percent.

The electoral fate of the extreme right was not much more enviable in the great democracies of England and Germany. In England, the United Kingdom Independence Party (UKIP), the heir to Jimmy Goldsmith's Referendum Party, tripled its score compared with the 1997 general elections and made its entry into the European Parliament in Strasbourg with three deputies. The political space ceded to the extreme right wing of the British National Party was very confined. The latter reached only 1.1 percent of the vote and found a significant audience only in several boroughs of East London (Dagenham and Barking). In Germany, the Republikaner, which had already failed in 1994 to reach its goal of getting more than 5 percent of the vote (it captured just 3.9 percent), eroded significantly with just 1.7 percent, while the NPD remained totally marginal, with only 0.4 percent. The Republikaner received more than 2 percent of the vote in only 4 of the 16 *länder* (Baden-Würtemberg, 3.3 percent; Saxony, 2.5 percent; Rhenish-Palatinate, 2.1 percent; Hesse, 2 percent). The episodic flare-ups of the extreme right in some depressed areas of former East Germany (for example, the 12.8 percent of the vote obtained in April 1988 by the DVU in the Landtag elections in Saxe-Anhalt) were not confirmed. In Saxe-Anhalt, the Republikaner party attracted only 1.3 percent of the vote, with the NPD gathering a very modest 0.7 percent.

In Italy, the extreme right's political space has shrunk substantially with the breakup of the MSI into two rival branches. The first, gathered around Gianfranco Fini, recognized that its participation in the Berlusconi government (1994) forced it to accelerate its democratic transmutation and go from the status of "excluded pole" of Italian political life to the status of being a component of the traditional Italian right, the Pole of Liberties. Compared with the postfascist wing that met with tremendous electoral success (15.7 percent of the vote in the 1996 legislative elections), the MSI-FT, which had remained loyal to the fascism of the Republic of Salò, was gradually marginalized. The 1.6 percent of the vote it won in the 1999 European elections was far from the 12.5 percent gathered by the MSI-AN in the European elections of 1994. The MSI-FT was able to obtain more than 2 percent of the vote only with great difficulty in a number of the country's center and southern regions, such as Latium (2.4 percent), Umbria (2.3), Marches (2.1), Abruzzi (2.3), Molise (2.4), and Apulia (2.3).*

There remain the two heavyweights of the extreme right wing in Europe: the National Front in France and the FPÖ in Austria. In France, the extreme right emerged greatly weakened following the breakup of the FN during the winter of 1998–1999 and the violent confrontation that opposed Jean-Marie Le Pen's National Front and the National Movement (MN) created by his former lieutenant Bruno Mégret. The FN had obtained 10.51 percent of the vote in 1994, while in June 1999 the two FN and MN fratricidal lists together drew just 9 percent (5.69 percent for the FN and 3.28 percent for the MN), not even enough to reach the threshold necessary for Parliamentary representation. At the heart of the reconstruction of the right following its gains in the 1998 regional elections, just one year later, the FN appeared to have become politically marginalized once again. Although, with about 9 percent of the vote, the extreme right's electorate still exists, it is clearly declining for the first time in 15 years. From 10.9 percent of the vote in 1984, 11.8 percent in 1989, and 10.5 percent in 1994, it declined to just 9 percent in 1999. Compared with the soaring gains of 1997–1998, in which the FN won 14.95 percent of the vote in the 1997 legislative elections and 15.01 percent in the 1998 regional contests, it is a very significant erosion indeed. Since the 1994 European elections, the extreme right has lost votes in most departments, barely managing to maintain its influence in only seven. This loss, in addition to the already mediocre electoral level it attained in 1994, was particularly noticeable in its Mediterranean bastions. Votes for the extreme right wing dropped by 4.9 percent in the Var department, 4.1 percent in the Alpes-Maritimes, and 3.2 percent in the Pyrénées-Orientales. The extreme right's political capital, already eroding significantly, saw its impact greatly weakened as a result of its internal divisions. Six percent of the vote plus 3 percent weigh significantly less than 9 percent.[3] The logic of the electoral breakup of the extreme right did not obey the same logic as the partisan breakup. When the FN split, a rather hasty analysis gave the dissident Bruno Mégret the advantage. In fact, his movement contributed greatly to decapitating the FN's party machinery, even if a majority of the movement's base appears to have remained loyal to Le Pen. To decapitate the party apparatus is one thing; convincing the electorate is another. This is the hard lesson of these elections for Mégret, who was unable to capture even one-sixth of the FN electorate. According to a postelection poll by the Sofres organization, of every 100 voters who chose the FN in the 1997 legislative elections and who voted in the European elections, only 17 percent chose the list headed by Mégret, 53 percent chose the list lead by Le Pen, 14 percent that headed by Charles Pasqua, 9 percent voted for the Hunting, Fishing, Nature and Tradition (CPNT) list, and 7 percent chose other parties. The majority of the FN electorate remained loyal to the old leader, with only one department, the Bouches-du-Rhône, where Mégret supporters control the two cities of Marignane and Vitrolles, giving Mégret a slight advantage. Le Pen's victory was important for him because his challenger, deprived of representation in the European Parliament in Strasbourg and confronted with real difficulties in assuring the material and political future of his little movement (now called the National Republican Movement), was politically marginalized. But this was a Pyrrhic victory. Enough "flesh" was removed from Le Pen's electoral base by the scission for his party, which until now seemed to have a political future, to discover the rough road of decline. For the first time in 15 years the FN found itself marginalized, with just 5.7 percent of the vote, behind even the Hunters' pressure group, which gathered 6.7 percent. Jean-Marie Le Pen was able to keep only over half of the traditional National Front electorate, the others having been attracted by the National Movement, Charles Pasqua's nationalism (his very good results in the Alpes-Maritimes and Var departments are the proof, with 20.4 percent and 19.2 percent of the vote respectively); or the Poujadism of the Hunters' list (in the Eure-et-Loir, Herault, and Somme departments, their electoral dynamism brought about a noticeable erosion of the extreme right). Finally, the increase in the number of abstentions particularly affected the FN's popular electorate, disoriented by the internal family divisions. It was among voters sympathetic to the FN that declared abstentions reached a height of 70 percent, according to a Sofres exit poll, compared with 45 percent among Rally for the Republic (RPR) sympathizers, 32 percent among Union for French Democracy (UDF) backers, 44 percent among Socialist Party supporters, and 36 percent among Communist Party voters. The electoral flourishing of the French extreme right seemed to have ended. The fishwives quarrel of winter 1998–1999 left its mark and made the FN lose its potential as a "party unlike the others." Mégret's strategy of rapprochement with other right-wing parties and his attempt to inoculate them with extreme right themes failed because the National Front's hard-core backers felt alien to Mégret's national-technocratic style.

In Austria, on the contrary, Jorg Haider has shown he knows how to keep together the two contradictory currents that led to the breakup of the National Front. On the one hand, the radical opposition, which relies on the new extreme right wing vote from workers (45 percent of workers voted for the FPÖ list), tried to maintain an equal distance between right- and left-wing parties and utilized rift as a strategy. On the other hand, a current of the party sought to be recognized by and integrated into the traditional right. Heir to the VDU (Union of Independents), which had brought together the pan-German conservative movements after World War II, the FPÖ, founded in 1956, had relatively mediocre electoral scores until the middle of the 1980s (4.98 percent of the vote in 1983).

The party, which united nationalists and liberals, participated in a government coalition with the Social Democrats (SPÖ) from 1983 to 1986. In December 1986, however, the liberal head of the party, Norbert Steger, was relegated to a minority position by the leader of the nationalist-populist wing, Jorg Haider. The SPÖ had denounced the coalition and provoked early elections in November. Jorg Haider's FPÖ experienced its first electoral breakthrough, gathering 9.73 percent of the vote. Since then, the FPÖ has continued to advance, from the 16.6 percent of the vote it attained in the 1990 legislative elections to 22.5 percent in 1994 and 21.9 percent in 1995. In the last legislative elections, held in October 1999, the party garnered 27.2 percent of the vote. The party remained well over the 20 percent mark in the 1996 and 1999 European elections, scoring 27.5 percent and 23.4 percent respectively. Even given the relatively low voter turnout for European elections, the FPÖ reached a very high percentage of the vote in its bastions of Carinthia (34.32 percent) and Vorarlberg (28.42 percent). The FPÖ's gaining of a strong foothold in the past 15 years parallels the attrition of an Austrian political system marked by a long coalition between Social Democrats and Christian Democrats as well as the dividing out of posts and privileges (the Proporz) between those parties since 1945. For a whole segment of the population, the FPÖ, given its outsider status with regard to the system, is in the best position to capture the protest vote against the coalition in power. In the 1999 European elections, this hostility toward the system was the leading motivation of the protest vote, even more important than hostility toward the European Union. This sense of rejection of a political system that was "blocked" flowered during the last legislative elections, when the FPÖ became the second most important Austrian party after the SPÖ and finished ahead of the conservatives of the Popular Party (ÖVP).

Despite the FPÖ's strong performance during the general elections, the party, like many other extreme right wing movements in Europe, also experienced an erosion in the June 1999 European elections. The FPÖ lost 4.1 percent of its vote between the elections of 1996 and 1999, while the various extreme right wing groupings in Germany lost 2.2 percent of their support between 1994 and 1999. Extreme right wing parties in French-speaking Belgium saw their votes decrease by 4.5 percent from 1994 to 1999. The only gains registered during this period for the extreme right were increases of 2.5 percent in Flemish-speaking Belgium (for the Vlaams Blok) and 3.6 percent in Denmark. What were the causes of this relative decline of the extreme right in Europe?

It is not surprising that national factors weigh heavily in the destinies of political forces that rely heavily on a logic of nationalism. In Austria, for example, the debate over whether to maintain the country's neutrality or join NATO hampered the FPÖ in the sense that the party, by favoring NATO membership, was not in touch on this issue with the great majority of Austrian public opinion. Questioned in June 1999, 81 percent of Austrians considered that maintaining their country's neutrality was a very important issue. In France, the spectacular breakup of the National Front during the winter of 1998–1999 contributed largely toward discrediting the extreme right in the June European elections. The public's decreasing concern with immigration as an issue also contributed to the decline of an extreme right wing that turned immigrants into scapegoats and made them the alpha and omega of its political campaigns. Finally, in Great Britain and France, the extreme right's political space was restricted by the vivacity of the competition from political parties running on anti-European and national sovereignty platforms (the UKIP in Great Britain and the RPF and CPNT in France).

Beyond these cyclical national factors, nevertheless, one of the common elements found in numerous countries is the pronounced proletarianization of the extreme right wing electorate. The percentage of workers who vote for the extreme right has increased strongly since the beginning of the 1990s, whether it be in the case of the National Front, the Republikaners, the DVU, VB, FRP, or the FPÖ. In France, workers represented 30 percent of National Front voters in the last legislative elections, in 1997.[4] In Belgium, they represented 36 percent of Vlaams Blok's electorate in 1991,[5] while in Denmark the percentage of workers who vote for the Progress Party has been higher than even that for the Social Democrats since 1988 (34 percent).[6] In Germany, working-class voters of the various extreme right wing parties (Rep, DVU, NPD) account for 43 percent of the party vote in the western *länder* and 69 percent in the eastern *länder.*[7] Finally, in Austria, the share of workers who voted for the FPÖ increased from 10 percent in 1986, before the Haider period, to 50 percent in 1996.[8] This proletarianization of the social bases of the extreme right has of course been accompanied by an increase in the low level of education of its voters. Sixty-four percent of extreme right wing voters in Germany have a poor level of education, compared with only 46 percent of the total population. In France, 80 percent of the National Front's electorate does not have a baccalaureate (Cevipof-Sofres poll taken in June 1997; Sofres exit poll in June 1999). The same is true in Belgium and Austria. These sociodemographic characteristics, as well as the male predominance of extreme right wing voters, are accompanied by a common value system on numerous issues: a strong hostility to immigrants and dissatisfaction with the political system and with democracy. These two aspects—of strong hostility toward foreigners, particularly immigrants from Third World countries, as well as rejecting the workings of democracy—are much more prevalent among extreme right wing voters than among voters of all other political parties. The FPÖ provides us with a relative exception to this rule. Despite Jorg Haider's xenophobic appeals, his electorate is less xenophobic than many other extreme right wing voters in Europe.[9] On the other hand, FPÖ voters are against the political system even more than their fellow voters in other European countries, with 51 percent saying they were motivated to vote for the FPÖ in the 1996 European elections by a desire to protest against the government and the traditional political parties. This was even more pronounced in 1999. In conclusion, although the success of extreme right wing parties appears linked to the more or less high level of satisfaction vis-à-vis the "system," the erosion of the European extreme right lies today in a decrease in concern about the economic crisis and discontent with political representation.

Generally speaking, the widespread perception of a profound economic and social crisis has eased. From the middle of the 1990s to the end of the decade, the impression that the economic situation was worsening diminished (table 10.3). Only 4 of the European Union's 15 countries are experiencing an opposite trend: Portugal, the Netherlands, Great Britain, and Denmark. In certain countries, such as Austria, Germany, Spain, and France, a relative upswing of optimism is quite evident. This return to a certain economic confidence could have contributed to the erosion of the German and French extreme right, which had turned to its advantage a deeply felt economic and social crisis in 1995 and 1996. Beyond the extreme right's capacity to exploit the uneasiness linked to the economic and social decline, the successes of the extreme-right European parties were due to a real capacity to politicize a certain sentiment of rejection of politics and discontent with democracy. Here also the "democratic uneasiness" appears to be undergoing a perceptible respite. While throughout the 1980s and until 1997, more or less 50 percent of European public opinion declared itself to be little

Table 10.3 Perception of the economic situation and the workings of democracy in European Union countries, 1994–1999

	Consider that the economic situation in the country in the coming year will be less favorable			Are fairly or wholly dissatisfied with the workings of democracy in the country (+ no answers)		
	Autumn 1995	Autumn 1998	Evolution	December 1994	March–April 1999	Evolution
All European countries	39	31	−8	51	39	−12
Germany	43	30	−13	35	28	−7
Austria	45	29	−16	–	36	–
Belgium	56	45	−11	41	51	+10
Denmark	18	41	+23	18	19	+1
Spain	30	10	−20	66	29	−37
Finland	20	18	−2	–	33	–
France	60	29	−31	44	41	−3
Greece	55	55	=	69	38	−31
Ireland	15	14	−1	31	26	−5
Italy	37	35	−2	73	66	−7
Luxembourg	28	21	−7	19	18	−1
Netherlands	26	35	+9	36	22	−14
Portugal	25	27	+2	52	43	−9
Great Britain	30	41	+11	49	36	−13
Sweden	29	22	−7	–	35	–

Source: Eurobarometer.

or not at all satisfied with the way democracy functions, since 1999 only 39 percent of those questioned in the 15 countries of the European Union share this sentiment (table 10.3). Only Denmark and Belgium, the two countries in which the extreme right progressed slightly in the last European elections, escape the trend. Everywhere else skepticism toward democracy is losing ground, at times quite significantly. It appears that, deprived in part of the motivating forces provided by economic and social protests on the one hand and political protests on the other, the extreme right in Europe was checked by the European elections in June 1999.

NOTES

1. Piero Ignazi, *L'estrema destra in Europa*, Il Mulino, 1994.
2. Xavier Casals I Meseguer, *La tentacion neofascista en Espana*, Barcelona: Plaza and Janès, 1998.
3. Annie Laurent, Pascal Perrineau, "L'extrême droite éclatée," *Revue Française de Science Politique*, vol. 49, no. 4–5, August–October 1999, pp. 663–641.
4. Pascal Perrineau, *Le symptôme Le Pen. Radiographie des électeurs du Front National*, Paris: Fayard, 1997, p. 210.
5. J. Billet and H. de Witte, "Attitudinal dispositions to vote for a new extreme right-wing party: The case of Vlaams Blok," *European Journal of Political Research*, 27, 1995, pp. 181–202.
6. J.G. Andersen and T. Bjorklund, *Radical right-wing populism in Scandinavia: From tax revolt to neo-liberalism and xenophobia*, forthcoming.
7. Jürgen W. Falter, *Wer wählt rechts?* Munich: Verlag C. H. Beck, 1994, pp. 99–100.
8. Patrick Moreau, "Le Freiheitliche Partei Österreich, parti national-libéral ou pulsion austro-fasciste?" *Pouvoirs*, no. 87, 1998, pp. 61–82.
9. O. Gabriel, "Rechtsextreme einstellungen in Europa: strucktur, entwicklung, verhaltenimplicationen," *Politische Vierteljahresschrift*, 37, 1996, pp. 344–360.

PART THREE

A Europe of Voters and Institutions

CHAPTER 11

Electoral Geography of Europe

Christian Vandermotten and Pablo Medina Lockhart

Contribution and Difficulty of a European Electoral Geography

On a national scale, electoral geography allows us to refine and spatialize the behaviors studied by electoral sociology, whose samples generally prove insufficiently representative at regional and local levels. It allows us to bring in additional hypotheses to explain the relationships between voting behaviors and the social, cultural, ideological, and economic features of the regions, even if it constantly requires being careful about improper interpretations of the correlations in terms of explanatory relationships. In the field of electoral geography, a relevant analysis always requires a good knowledge of the overall historical and social background.

The temptation is strong to extend this reasoning to all of the European electoral results (for instance the 1999 European Parliament election), insofar as we are provided with a synchronous (which, of course, does not prevent allowing for the moments that those elections took place within national events and political cycles) and presumably homogeneous view of the electoral behaviors on the global European scale.[1] All this involves increased difficulties in criticism and interpretation, due to at least two factors.

First, the degree and the mode of approval of the European vision by the national electorates and the rates of abstention (or of null votes) are quite different according to the countries, so that the results of the European elections can be more or less biased in comparison with the national legislative elections, the results of which appear as more representative of the national sociopolitical spectra.[2] On the contrary, less emphasis on efficient voting as well as proportional representation may lead to a more open expression of the political attitudes at the European elections in the countries with majority representation. The second—and not smaller—difficulty lies in the necessity to elaborate a reasoned classification of the parties that would be not only relevant but simple enough to be used as a basis for a global cartography.

Though not uninteresting, the groupings of the parties into political groups at the Strasbourg Parliament do not make up a sufficient basis, as such, on which to found a classification (see table 11.1). The European party system still has to be further defined and structured. Some groupings may be sheer technical confederations, and 26 deputies out of 626 remain unregistered. Additionally, there is the question of how to classify the parties that are not represented in this assembly.

In our opinion, classifications based upon the theory of the cleavages and inspired by the works of Seymour Lipset and Stein Rokkan (1967) or Daniel Seiler (1980) are not operational enough. Such theories are indeed more appropriate to explain the origins of the historical formation of the party systems within their own national contexts—and even so, from this point of view they can be criticized—than to work out a current classification in a cross-national context. Parties closely positioned on the classification axes of Rokkan and Seiler can have quite divergent views on a number of present political stakes, and more particularly the European construction.

From this point of view, a classification based exclusively upon the positioning of the party programs on a left–right axis can also prove completely inappropriate.[3] Contrasted attitudes toward the European construction concept can in fact not only separate parties relatively close to each other in some matters on the left–right axis, but can also be found within certain parties such as the British Labour and Conservative Parties. Paradoxically then, it would not be possible to establish a global cartography that reflects the geography of the intensity or the forms of the agreement with the European vision.

The electoral maps as they appear hereafter are based upon a classification that has to be seen as the result of a simplifying compromise, sometimes contestable, but anyway permitting an outline of an electoral spatiality of Europe. Classification of parties is essentially based upon a reading of their programs and structures, but in case of doubt one can take into account—though not systematically—their belonging to the groups of the Strasbourg Parliament or to international partisan organizations such as the Socialist International or the European People's Party. The classification is based upon a left-right axis, which remains in every aspect the most meaningful when analyzing electoral behaviors. We suggest a rough classification into three groups: left; center and center right; and reactionary and extreme right—obviously more manageable than the division into 11 political families

Table 11.1 The new lineup in the European Parliament after the 1999 elections

	European Unitarian Left/ Nordic Greens	European Socialist Party	Greens/ European Free Alliance	European Democrat and Reform Liberals	European People's Party/ European Democrats	Europe of Democracies and Diversities	Union for Europe of the Nations	Unregistered	Total
Extreme left	5							1 (1)*	6 (1)*
Communists	35								35
Socialists	2	180 (1)*	3 (3)*			1			186 (4)*
Ecologists			38						38
Left	42	180 (1)*	41 (3)*			1		1 (1)*	265 (5)*
Liberal and Conservative centrists			7 (7)*	45 (4)*	135 (1)*	6	6	7	206 (12)*
Christian Democrats					97 (1)*				97 (1)*
Agrarians				5					5
Center right			7 (7)*	50 (4)*	232 (2)*	6	6	7	308 (13)*
Populists and Poujadists				1 (1)*	1	6	24	9 (3)*	41 (3)*
Authoritarian extreme right								8 (2)*	8 (2)*
Religious extreme right						3		1	4
Extreme right				1 (1)*	1	9	24	18 (5)*	53 (6)*
Total	42	180 (1)*	48 (10)*	51 (5)*	233 (2)*	16	30	26 (6)*	626 (24)*

* The columns show the makeup of the political groups in Strasbourg, while the lines represent the groupings into political families as they will be used in the present study. The numbers of seats held by the regionalist parties are shown in parentheses.

Table 11.2 Results of the 1999 European elections by political family* (in percentage of the valid votes)

	Extreme left and Communists	Socialists and Labor	Ecologists	Total left	Liberals and Conservatives	Christian Democrats	Agrarians	Total center and classical right	Reactionary, populist or poujadist right	Authoritarian extreme right	Total extreme right	Others	Of which all regionalist parties
Austria	0.7	31.7	9.3	41.7	2.7	32.2	–	34.9	23.4	–	23.4	–	–
of which Vienna	1.5	34.3	15.0	50.9	5.5	21.7	–	27.2	21.9	–	21.9	–	–
Germany	6.2	30.6	7.5	44.3	3.0	48.7	–	51.7	2.6	0.4	3.0	0.9	0.1
of which former West Germany without Berlin	1.5	33.0	8.4	42.9	3.4	50.2	–	53.6	2.4	0.3	2.7	0.8	0.1
of which northern and central former West Germany[a]	1.6	36.7	8.1	46.4	3.5	47.0	–	50.5	2.0	0.3	2.3	0.8	–
of which southern former West Germany[b]	1.0	23.6	9.4	34.1	3.2	58.3	–	61.5	3.4	0.2	3.6	0.8	0.2
of which former GDR and Berlin	22.8	24.3	5.3	52.4	2.1	40.3	–	42.4	2.9	0.8	3.7	1.4	–
Switzerland	2.0	22.7	5.0	29.6	22.9	17.6	–	40.5	27.6	–	27.6	2.3	–
Luxembourg	2.7	23.2	10.5	36.4	28.9	31.1	–	60.0	1.8	–	1.8	–	–
Belgium	1.5	18.7	15.9	36.1	31.4	18.7	–	50.1	0.4	11.4	11.8	2.0	17.1
of which Flanders	0.6	14.3	11.9	26.8	34.5	21.8	–	56.3	0.4	14.8	15.2	1.7	27.1
of which Brussels-Capital	2.1	17.2	26.1	45.4	33.7	10.7	–	44.5	0.4	7.8	8.1	2.0	4.3
of which Wallonia	3.3	27.9	21.5	52.7	24.8	14.2	–	39.0	0.5	5.4	5.8	2.4	–
Netherlands	4.9	20.0	11.6	36.5	25.4	27.5	–	52.9	10.3	–	10.3	0.3	–

of which Randstad Holland^c	5.9	24.6	16.3	46.7	17.4	23.4	—	40.8	11.9	—	11.9	0.6	—
United Kingdom	2.0	30.8	5.9	38.7	57.4	—	—	57.4	1.8	1.0	2.8	1.1	7.1
of which England	1.1	29.0	6.5	36.5	61.3	—	—	61.3	—	1.2	1.2	1.0	—
of which southern England^d	0.9	25.0	7.4	33.3	65.0	—	—	65.0	—	1.1	1.1	0.6	—
of which Greater London	1.8	35.0	7.7	44.5	52.7	—	—	52.7	—	1.6	1.6	1.3	—
of which central and northern England	1.2	32.9	5.5	39.7	57.7	—	—	57.7	—	1.3	1.3	1.3	—
of which Wales	0.7	31.9	2.6	35.1	64.6	—	—	64.6	—	—	—	0.3	29.6
of which Scotland	0.9	55.9	5.8	62.6	32.6	—	—	32.6	0.2	0.4	0.5	4.2	27.2
of which Northern Ireland	17.3	28.1	—	45.4	26.0	—	—	26.0	28.4	—	28.4	0.1	45.4
Ireland	9.0	11.4	6.7	27.1	71.6	—	—	71.6	—	—	—	—	—
Denmark	7.4	24.3	—	31.7	59.9	2.0	—	61.9	5.6	0.8	6.4	—	—
Sweden	15.8	26.1	9.4	51.3	34.4	7.7	6.0	48.0	—	—	—	0.6	—
Finland	10.1	17.9	16.2	44.3	32.0	2.5	20.5	55.0	—	0.8	0.8	—	6.0
Norway	7.8	35.0	0.2	43.0	19.1	13.7	7.9	40.7	15.9	0.1	16.1	0.2	0.4
Iceland	—	26.8	9.1	35.9	44.9	—	18.4	63.3	—	—	—	—	—
France	12.1	22.9	11.3	46.2	21.8	—	—	21.8	21.8	9.1	30.8	1.1	—
of which Ile-de-France	13.3	21.5	14.4	49.2	23.4	—	—	23.4	17.0	9.5	26.5	0.9	—
Spain	7.5	37.6	2.1	47.2	52.2	—	—	52.2	—	0.2	0.2	0.4	14.8
of which north, northwest, center and east^f	6.1	37.3	1.8	45.2	54.2	—	—	54.2	—	0.3	0.3	0.4	8.4
of which Pays Basque	24.0	19.9	0.6	44.5	55.0	—	—	55.0	—	0.1	0.1	0.4	54.9
of which Asturias	10.8	42.2	1.0	54.1	45.6	—	—	45.6	—	0.1	0.1	0.2	5.5
of which Catalonia	3.0	35.8	7.0	45.8	53.2	—	—	53.2	—	0.2	0.2	0.8	36.6
of which south, except the Canaries^g	9.0	43.9	0.8	53.7	45.8	—	—	45.8	—	0.2	0.2	0.3	5.1
Portugal	13.6	44.6	0.4	58.6	32.2	—	—	32.2	8.5	—	8.5	0.7	—
of which Lisbon and south of Tagus	21.9	44.9	0.5	67.4	25.0	—	—	25.0	6.8	—	6.8	0.8	—

Table 11.2 (Continued)

	Extreme left and Communists	Socialists and Labor	Ecologists	Total left	Liberals and Conservatives	Christian Democrats	Agrarians	Total center and classical right	Reactionary, populist or poujadist right	Authoritarian extreme right	Total extreme right	Others	Of which all regionalist parties
of which north of Tagus	*7.8*	*44.7*	*0.3*	*52.9*	*36.8*	–	–	*36.8*	*9.6*	–	*9.6*	*0.7*	–
Italy	5.9	19.7	1.7	27.3	42.7	10.2	–	52.9	17.4	2.4	19.7	0.1	6.4
of which north, except Upper Adige	*5.4*	*14.2*	*1.8*	*21.4*	*46.3*	*7.1*	–	*53.4*	*19.4*	*1.7*	*21.1*	–	*11.4*
of which Upper Adige	*1.3*	*3.7*	*6.7*	*11.7*	*17.1*	*57.6*	–	*74.7*	*12.7*	*1.0*	*13.7*	*0.1*	*62.9*
of which center[h]	*7.4*	*28.5*	*1.7*	*37.6*	*30.8*	*8.8*	–	*39.6*	*19.0*	*3.7*	*22.7*	–	*2.3*
of which south	*5.2*	*18.2*	*1.5*	*24.9*	*47.3*	*13.8*	–	*61.1*	*12.1*	*1.9*	*14.0*	–	*1.1*
Greece	13.1	40.2	–	53.3	40.7	–	–	40.7	–	–	–	6.0	–
of which Attica	*17.9*	*36.3*	–	*54.2*	*37.4*	–	–	*37.4*	–	–	–	*8.4*	–
Europe	6.8	28.1	6.5	41.5	30.7	15.9	0.5	47.1	8.2	2.3	10.4	0.9	3.7

*Legislative elections of 1997 in Norway, and of 1999 in Iceland and Switzerland.
[a]All former länder, except West Berlin, Bavaria, and Baden-Württemberg.
[b]Bavaria and Bade-Würtemberg.
[c]Provinces of Noord- and Zuid-Holland and of Utrecht.
[d]Southwest, Southeast, and East Anglia regions.
[e]Tyne-and-Wear, West Yorkshire, Greater Manchester, Merseyside, West Midlands.
[f]Except the Asturias, the Pays Basque, Catalonia.
[g]Extremadura, Andalusia, Castile-Mancha, Murcia.
[h]Emilia-Romagna, Tuscany, Umbria, the Marches, Latium.

(grouped into 9 in our cartography) that permits refining of them. The national results, as well as those of larger areas with very marked political attitudes, have been summarized in this framework (see table 11.2).

In addition to the scores of the main families, we present in map 1 (see the Appendices) the results of the regionalist parties on the whole, whatever their position on the left–right axis may be. The criteria we have adopted in order to define the boundaries between the regionalist parties and the others is an electoral presence limited to a part of a national territory and a strong positioning in favor of the confirmation or the strengthening of an acquired regional autonomy, or even a more radical contestation of the existing state structure in the direction of an increased regional autonomy, or, naturally, independence.[4] In this way it is obvious that we do not consider the Socialist, Liberal, and Christian Democratic families in Belgium as belonging to the regionalist parties, even if they are divided according to a linguistic split. Among the regionalist parties, we will include the Northern Irish parties claiming incorporation into the Republic of Ireland (Sinn Féin) or a deepening of the provincial autonomy on a constitutional basis (the SDLP), but not those boosting the preservation of a close alliance with Great Britain, even when their extremist positions lead them to want to reduce the rights of the Catholic minority. Meanwhile, Sinn Féin will not appear among the regionalist parties in the Republic of Ireland, as it would entail, on the contrary, an enlargement of the state by incorporation of Northern Ireland. We do not categorize as regionalist the parties that do not bring into question the state organizational structures at regional level or that do not center their platforms on the defense of the acquired autonomy, even if their electoral grounds are strongly localized in a determined region, the best illustrations of this being the German PDS and CSU.

The Classification of the European Political Parties

The Left (41.5 percent of the Votes on European Scale[5])

It is in this group that the classification turns out to be the easiest on the whole. It includes four families, but we will group the extreme left and the Communists.

The extreme left and the communists (6.8 percent). The extreme left family is composed of parties that claim to belong to the working-class movement and are situated to the left of the Socialist movement, but are devoid of historical links with the Communist movement such as it used to be organized under the leadership of the Soviet Union. The only two parties clearly bound to this family and having representatives in the Strasbourg Parliament are, in France, Lutte Ouvrière and the Ligue Communiste Révolutionnaire, together on the list led by Arlette Laguiller[6] (5.2 percent of the votes in France). The five representatives elected by the LO-LCR have joined the six of the PCF and their allies in the group of the European Unitarian Left/Scandinavian Greens. We have included in the extreme left some regionalist parties with particularly radical socioeconomic programs, even if they do not share the internationalist ideology, a pillar of the extreme left, namely Euskal Herritarrok (EH) (19.0 percent of the votes in the Pays Basque and in Navarre, one unregistered representative in Strasbourg) and Sinn Féin (17.3 percent of the votes in Northern Ireland and 8.3 percent in the Republic of Ireland).

The parties making up the Communist family are situated in the prolonging of the heritage of the international Communist movement, independently of their critical repositioning or their opening to the European vision and to alliances with left-wing circles of other affiliation. The Italian Left Democrats (DS, former PDS) have nevertheless been included in the Socialist family as it had joined the Socialist International. Although their opinions may diverge on subjects of importance, their common heritage allows those parties to form a coherent group in Strasbourg, in which we also find the German PDS (21.8 percent of the votes in the former GDR, including Berlin), the PCF (6.8 percent of the votes in France, with the associated independents), the two Greek Communist Parties (respectively 8.2 percent for the KKE and 4.9 percent for the pro-European SIN), the two Italian parties that maintained a Communist reference (respectively 3.9 percent for the RC and 2.0 percent for the PCI), and the alliances based upon the communist movement in Spain (5.9 percent for the IU-EUA) and in Portugal (10.7 percent of the votes for the CDU). In the Scandinavian countries, except maybe in Finland, as well as in the Netherlands, the affiliation with the traditional Communist movement is not so obvious within the socialist-popular unions, situated to the left of Social Democracy and showing a marked ecologist and pacifist component: the Danish SF (7.4 percent), the Swedish Vp (15.8 percent), the Finnish VAS (9.5),[7] and the Dutch SP (4.8). We have considered that the ecologist component is dominant among the Dutch Groenlinks even if former Communists are present there too.[8]

The social democrats (28.1). The Socialist family too makes up a coherent group in Strasbourg, including all the parties belonging to the Socialist Internationale. Despite this coherence, built on a common reformist practice, the Socialist and Social Democratic Parties can be attached to a diversity of political and organizational traditions. The German (30.6 percent of the votes), Austrian (31.7 percent), Swedish (26.1), Belgian (18.7), Luxembourg (24.2) and Finnish (18.0) parties for instance belong to the Social Democratic tradition, with a strong structuration, in fact a bureaucratization, of the party organization. In the British labor movement (28.0 percent of the votes, including the Northern Irish SDLP), to which the Irish (11.4 percent, the lowest score of all the EU countries), Dutch (20.0 percent), and even Danish (17.1 percent) parties can be linked, the weight of ideology is very limited. These parties traditionally had a low rate of organization but were greatly influenced by the trade union movement, from which they are now separate. They give little importance to egalitarian views. The Spanish (35.9 percent of the votes), French (21.9 percent, allied with the MRG and MDC), and Portuguese (44.6 percent, the highest score of the EU) parties, as well as the Italian Social Democracy (2.1 percent) (thus to the difference of the DS, 17.4 percent), are rather close to a Socialist and republican tradition, founded on an egalitarian and generous discourse but traditionally little organized. The Greek PASOK (33.5 percent of the votes) is also to be situated in a republican tradition of center left, with some very nationalist accents.[9] These Social Democratic parties have moved toward a center left practice, stressing their reformist nature and often developing, beyond a modernist discourse, the features of catchall parties, appealing to self-employed workers and executives alike, as well as to the Christians in the Catholic countries where the Socialist movement used to be deeply bound to the struggle for secularism, as in Belgium. In Britain, some trends of Labour have become more and more similar to the Christian Democracy in countries that have long Catholic traditions. Overall, the cohesion of the group at the Strasbourg Parliament is also based on a large approval of the European vision, except for its deepening, as is the case of certain British Labour Party members.

As in the Communist tradition, there are no organized regionalist trends within the European Socialist group, if we except the Northern Irish Social Democratic representative, who claims devolution of constitutional power for Northern Ireland, an objective shared by the British Labour Party itself. Except for some small groups that are not represented in the European Parliament, a few parties linked to the Social Democratic tradition by their socioeconomic views do not belong to the ESP group in the European Parliament. This is the case of the Greek dissidents of the Democrat Party (DIKKI), which has joined the Communist group in Strasbourg (6.7 percent of the votes in Greece); the strongly anti-European Danish Social Democratic wing (Folkebevaegelsen mod EU, 7.2 percent); and some regionalist parties whose economic and social programs are clearly center left or left oriented (the Scottish SNP, 27.2 percent; and the BNG, 22.6 percent of the votes in Galicia).

The ecologists (6.5). If certain segments of political environmentalists still refuse to position themselves to the left or to the right, it seems nevertheless obvious that a large majority of the ecologist electorate is, from a sociological point of view, positioned to the left. As to the ecologist parties that are represented in the European Parliament, their left-wing positioning is still more evident. Meanwhile, a good deal of their electorate lacks working-class tradition and comes from middle-class or wealthy intellectual origins, or from the Christian left. Furthermore, the ecologist parties position themselves on a modernist/postmodernist cleavage, which leads them to relativize the Promethean values, as they are conveyed as much by the traditional left as by the liberal right, assimilating technological advancement, economic growth, and social progress. All this results in geographies differing quite a lot from the classical left.

The Greens/European Free Alliance group in Strasbourg includes the German (6.5 percent) and Austrian (9.3 percent) Grünen, the Swedish Environment Party (9.4 percent), the Belgian ecologists (Agalev with 11.9 percent of the Flemish votes; Ecolo, 21.5 percent of the Walloon votes; and, together, 26.1 percent of the votes in Brussels), the Greens of Luxembourg (10.5 percent), and the British (5.9), Irish (6.7), Finnish (13.4), French (9.8) and Italian (1.7)[10] ecologist parties. A left-wing positioning is especially explicit among the Dutch Groen links (11.6).[11] In Portugal and, outside the EU, in Norway, political environmentalism is associated with far-left alliances. Some very small other ecologist or centrist movements campaigning for animal rights, with no representatives at all in Strasbourg, have been included in the results of the ecologist families, except the French Génération Ecologie, electorally bound to the RPR/DL. Those parties seldom win more than 1 percent of the votes on the national scale.

Let us remind the readers that the Greens have associated with ten left-wing regionalist representatives stemming either from the left (the Scottish SNP, the Galician BNG) or from the cross-class center (the Flemish Volksunie, the Welsh Plaid Cymru, certain Andalusian and Basque regionalists), with no implication of real political common stands.

The Center, the Center Right, and the Classical Right (47.1 percent of the Votes on European Scale)

To this group belong the center and Conservative parties, the programs of which globally agree with the working of the Establishment, of the state, and generally of the EU, at least to the point that the latter is seen by the major economic forces as necessary to their optimal expansion.[12] An indisputable classification is hard to achieve within this group, where most parties belong to the EPP (233 representatives) and the ELDR (51 representatives) in the European Parliament, presenting a difficulty that keeps growing as time goes by. In the past indeed, the church-state cleavage separated rather clearly the parties of the radical liberal family from those of the Christian Democratic family and prohibited Christian left-wing circles from joining parties of the classical left, at least in the Catholic countries. Such a cleavage tends to be less today, considering the growing secularization of political life, the prevailing neoliberal assertion, and the increasing catchall character of the center and center right parties (as well as some Social Democratic parties). The growing difficulty in distributing the parties within the center and center right groups is confirmed by the makeup of the EPP group in the Strasbourg Parliament, where there are as many parties with clearly Christian Democratic references as Conservative parties, even if the opposite is not true: There are no Christian Democratic parties affiliated with the EDRL group. Yet one cannot maintain that the latter is further to the left than the EPP group as far as the socioeconomic positions of the majority of its representatives are concerned.

Despite these difficulties in subdividing the center and the center right, and beyond the affiliations expressed in Strasbourg, we will stick to a classification into three families: a Christian Democratic family, a classical centrist/liberal/conservative family (with a large spectrum of socioeconomic views), and a more residual agrarian family.

The christian democrats (15.9). The Christian Democracy, from a historical point of view, has developed from the defense of Christian and family values and positions in relation to church-state cleavage. The Christian Democratic Parties now address a larger social spectrum. They are characterized by an organization in pillars, the existence of the Christian trade union movement, the organization of the rural class, etc. They do not contest the primacy of the state over the church, but they assert that legislation has to draw its inspiration from Christian values. With the weakening of the church-state cleavage, the Christian Democracy generally succeeds less in maintaining common positions between its bourgeois, middle-class, or rural traditional conservative circles and the Christian working class. It remains nevertheless very present in median Europe,[13] in spite of its weakening in Belgium due to the Liberal growth—48.7 percent of the electors in Germany[14] and 41.5 percent in the area comprising Benelux, Germany, Austria, and Switzerland. In Italy, however, the Christian Democracy has imploded, largely to the benefit of the classical liberal right of Forza Italia,[15] retaining only 10.2 percent of the votes, split up among three parties to which one can add two regionalist parties with a strong Christian Democratic trend: the Süd-Tyroler Volkspartei and the Union Valdôtaine.

The Christian Democratic parties as such are absent from a number of countries: Orthodox Greece; some Protestant states with no church-state cleavage; Ireland, where the assertion of a religious identity has given birth to the nation; France, where state tradition and republican legitimacy are especially strong; as well as Spain and Portugal. In these latter two countries, indeed, the recent character of the democratic party system and the links binding the church to the fascist power made any organic and explicit association between the working-class forces, even Catholic, and conservative circles impossible. Yet regular religious observances in Spain as in Portugal are always superior among the electors of the center and center right parties than among those of the left.

Although they make up part of Protestant countries, some small Scandinavian Lutheran parties are included in the social Christian family. While they strongly assert their attachment to the traditional Christian values and blame the permissiveness of the Establishment, they do not dispute the state's legitimacy in relation to religious dogmas. These are: in Finland, the Finnish Christian League (SKL, 2.5 percent); in Sweden, the Christian Democratic Community Party (KDS, 7.7 percent); in Denmark, the Popular Christian Party (KrF, 2.0 percent); and in Norway, the Popular Christian Party (KRF, 13.7 percent at the latest legislative elections).[16]

The centrists, classical liberals and conservatives (30.7). These parties dominate the center and center right almost exclusively in the countries outside median Europe. The wing historically born of the liberal-radical, anticlerical tradition makes up the core of the EDRL group in the European Parliament, while the parties with a Conservative tradition are more often members of the EPP group, together with the Christian Democrats. However, some liberal parties born of the church-state cleavage have now opened themselves to the Christian circles, more or less successfully, and, considering the strengthening of the neoliberal positions, such a grouping is no longer very significant in terms of left-right cleavage. Many representatives of the social Christian wing in the EPP share socioeconomic stands further to the left than those of some Liberal representatives belonging to the EDRL. As to the Conservative parties, if they are generally very close to the Establishment, with which they share the fundamental values in terms of national and international liberalization of the economy, their catchall character often incorporates populist trends appealing to an overcautious middle class, staggered or threatened by the economic globalization and the insecurity it brings. So the limit is not always easy to mark between certain populist tendencies that we will classify hereafter in the group of the reactionary right.

The Agrarians (0.5; 10.6 in Scandinavia except Denmark). The agrarians represent a variant of the centrist populism. They are limited to the

Table 11.3 Centrist, classical liberal, and conservative/agrarian parties in the Strasbourg Parliament.*

	Social–Liberal, Liberal–Radical and Liberal affiliation and references	Agrarian affiliation and references	Conservative affiliation and references	Others
Switzerland	Radical Democratic Party (19.9%) Swiss Liberal Party (2.3%) Independents' Alliance (0.7%)			
Luxembourg	Democratic Party (20.1%)			
Belgium	VLD (Flemish Liberal-Democrats) (13.8%) Liberal Reformer Party(a) (9.8%)			[Flemish Volksunie-ID21] 7.7%
Netherlands	VVD (Popular Party for Freedom and Democracy) (19.2%) D66 (Democrats 66) (5.6%)			
United Kingdom	Liberal Democrats (11.9%)		Conservatives (33.5%) [UK Independence Party] (6.5%) Ulster Unionist Party (1.1%)	[Playd Cymru] (1.7%)
Ireland			Fine Gael(b) (23.7%) [Fianna Fail] (39.8%)	
Denmark	Venstre (= the Left) (22.1%) Radikale Venstre (9.1%)		Conservative Popular Party (8.4%) [JuniBevaegelsen mod Union] (16.7%)	
Sweden	Popular Liberal Party (FpL) (13.8%)	Party of the Center (6%)	Unified Moderate Party (20.6%)	
Finland		KESK (Finnish Center) (20.5%)	National Rally (KOK) (26%)	Sweden's Popular Party RKP-SFP (6%)

Table 11.3 Continued

	Social-Liberal, Liberal-Radical and Liberal affiliation and references	Agrarian affiliation and references	Conservative affiliation and references	Others
Norway	Venstre (= the Left) (4.5%)	Sp (Party of the Center) (7.9%)	Höyre (= the Right) (14.3%)	Party of the Maritime Areas (0.4%)
Iceland	FF (4.2%)	FSF (Party of Progress (18.4%)	SFF (Independence Party) (40%)	
France	UDF (9.2%)		RPR-DL (12.5%)	
Spain			Popular Party (PP) (40.4%)	Coalicion Europea (3.3%) Convergencia i Union (4.5%)[d] [Coalicion Nacionalista Europa de los Pueblos] (3%)
Portugal			Social Democrat Party (PSD) (32.2%)	
Italy	Republican Party/Liberals (PRI/Lib) (0.5%) Italian Renovation (RI/Dini) (1.1%) I Democratici/Prodi (7.7%) [Bonino list] (7.7%) UnioneDemocratici Europei (1.6%)		Forza Italia (23.3%)	
Greece			New Democracy (ND) (36.7%)	

* (as well as in the national parliaments as concerns the three non-EU countries)
The percentage of national votes are in parentheses. Underlined are the EPP members in Strasbourg, not underlined are the EDRL members, and in brackets are the parties that do not belong to either of these two groups.

Notes: [a] Includes in Brussels the French-Speakers' Front and the MCC, dissidents of social Christian origin.
[b] Seen by the electors as more liberal than Fianna Fail. To this result must be added the 8.1 percent of the votes won by the independents, who gained two representatives, one of whom has joined the EPP group and the second the EDRL.
[c] The Canary Islands' representative has become a member of the Liberal group in Strasbourg, while the Andalusian has joined the Greens/European Free Alliance group.
[d] The two representatives of Catalonia's Democratic Convergence are affiliated with the Liberal group in Strasbourg, but the deputy of Catalonia's Democratic Union has joined the EPP group.

Scandinavian countries, where their history is based both upon the democratic community traditions of conservative and Protestant peasant classes and upon the protection needs of agriculture in their few fertile regions. It is undoubtedly no accident that these countries have remained outside the EU for a long time—some of them still are. The agrarians are the counterpart, in the Protestant rural regions of Scandinavia, of the Christian Democracy in the Catholic rural areas. The weakening of their social bases has led them to a reconversion into center-right parties, which now also try to attract the townspeople.

The Reactionary Right and the Extreme Right
(10.4 percent of the Votes in European Scale)

We will make a distinction between this third group and the center and center right, including the classical right. The parties of this group advocate an ideology of withdrawal and call into question, in a reactionary[17] or authoritarian way, the political system or at least the main elements of the dominant economic and political trends or structures. In this group we will distinguish three families: the populist and/or Poujadist reactionary right, the religious extreme right, and the authoritarian extreme right. The first two will be dealt with simultaneously, given the very low representativity of the second.

The reactionary populist and/or Poujadist right and the religious extreme right (8.2). Even if they do not always share radical extremist political views or challenge the democratic system, the parties and the electorate of the reactionary right differ from those of the classical right by a marked identity withdrawal and a more or less explicit refusal of the consequences of economic liberalization and globalization on the part of middle class or self-employed circles, or even of a fraction of the working class lacking the traditional socialization forms of the worker movement, or having lost them by reason of aging or unemployment. One can distinguish five subfamilies here without being able to operationalize this new subdivision at the level of the classification of the parties and probably less still as far as the underlying attitudes of their electorates are concerned.

The first subfamily includes the parties born of the classical right, to which they sometimes remain very close but from which they nevertheless differ in a marked reluctance to lose their national identity and in a very restrictive view of the European construction, which they see as a threat to the economic and social positions of their electorate, undermined by economic globalization. Their reluctance to deepen the European vision is accompanied by a very strong identification with the national concept. The most important list expressing such a trend is the French RPFIE

(13.1 percent of the national votes), to which can be added the Portuguese PP/CDS (8.5 percent) and the Dansk Folkepartei (DF) (5.6 percent of the votes in Denmark). But a number of parties we have classed in the classical center right, such as the U.K. Independence Party (6.5) or even large segments of the British Conservatives, are rather close to the positions of this subfamily, as are the Danish anti-European center right (16.7 percent for JuniBevaegelsen mod Union) and probably the Irish Fianna Fail (39.8), although in Ireland the party system is hard to interpret and the adherence to the European vision is considerable. The Austrian FPÖ (23.4) has moved toward a populist right-wing position on the edge of the authoritarian extreme right after drifting from being a standard liberal party, which took on, it is true, the heritage of a pro-German authoritarian liberal–national trend and had welcomed plenty of "little Nazis." In the same way, outside the EU, the Swiss UDC/SVP (22.6) has moved from an agrarian affiliation toward radical populist right-wing positions, also on the border of the extreme right, but without developing the centralizing argument, unknown to the Swiss tradition, or the leader cult.[18]

The positions of the second subfamily are pretty similar to those of the preceding, in particular the strong identification to with the concept of national state but on the basis of a historical positioning in the authoritarian extreme right. This is the case of the National Alliance (11.6 percent of the Italian votes won in union with the Segni list), a historical product of the Italian Fascist movement, and of some Salazarist trends of the Portuguese PP/CDS. On the edge of the authoritarian extreme right we also mention the German "Republicans" (1.7 percent of the German electorate, 2.5 percent in Bavaria and Baden-Würtemberg, no representatives in Strasbourg), who express a number of currents issued from the extreme conservatism of Catholic southern Germany rather than an obvious nostalgia for Nazism.

These first two subfamilies are both members within the European Parliament of the Union for Europe of the Nations group, which also includes the Irish Fianna Fail. As mentioned above, and though they appear as more centrist in other respects, a number of very anti-European movements or parties could be assimilated to these subfamilies, such as the Danish JB and the British UKIP, which belong to the Europe of Democracies and Diversities group.

The third subfamily shares the positions of the first one, except that its identity (not "xenophobic") withdrawal is expressed here in a questioning of the national state, in favor of a regional withdrawal in countries with major internal economic disparities, or where the national unity is slow in coming or else the shaping of the nation-state is not really completed. These views are explicitly expressed by the Northern Italian League (5.1 percent of the votes, but 10.1 percent in the north, including Emilia-Romagna but excluding Upper Adige). In Flanders, a part of the electorate of the Vlaams Blok, a party that we will hereafter relate to the authoritarian extreme right on the basis of its program, is probably to be included in this category. The fourth subfamily has a more explicitly Poujadist character, as expressed through its antistate, antitax, and antiregulations program. The fifth, very close to the preceding, concentrates its rejection of structures, taxation, and regulations on one specific field, such as the defense of car drivers, of hunters, or other particular categories. The only two movements of this category represented in Strasbourg are the French list Chasse, Pêche, Nature et Traditions, with an amazing score of 6.9 percent of the national votes, and the Italian Pensioners Party (0.7 percent of the votes and one representative member of the EPP).

The family of the religious extreme right includes small Protestant groups asserting that the state has to submit to religious values, which are not democratically disputable, and that the positions adopted by the official religious authorities are not rigorous enough. In this category we can include three Dutch Protestant parties (RPF, SGP, and GPV, totaling 9.5 percent of the Dutch electorate and three representatives in Strasbourg).[19] The Northern Irish Democratic Unionist Party (28.4 percent in Northern Ireland) is at the same time an extremist Protestant party and a defender of the strengthening of state authority, demanding total union with Britain, which is seen as a bastion against the Catholics. In spite of their quite moralistic religious positions, the Scandinavian small Protestant parties are rather similar to a Protestant version of the Christian Democratic family. Besides, they have joined the EPP group in Strasbourg.

The authoritarian extreme right (2.3). The parties of the authoritarian extreme right are clearly supporters of traditions of law and order and the leader cult. They nourish a xenophobic or racist rejection of foreigners, reject some of the democratic state structures, claim the promotion of a strong national state (or of a state to be created from the decay of a state, as is the case of the Vlaams Blok in Belgium), feel a strong distrust of big "cosmopolitan" capitalism, and are not opposed to Fascist or Nazi theories, even if they do not explicitly claim to adhere to them, which is electorally less and less profitable. In this category the Flemish Vlaams Blok (14.8 percent in Flanders), the French National Front (5.7 percent or, if we include the dissident movement of Bruno Mégret, 9.0 percent), and the Tricolor Italian Social Flame movement (MS-FT, 2.4 percent) are represented in Strasbourg. The representatives of the authoritarian extreme right do not belong to any group in Strasbourg.

Electoral Geography of the Political Families

The comment of the electoral maps will be globalized by groups of political families.

The Left

If the map of the left-wing voting appears rather homogeneous throughout the European space (map 2—see the Appendices), it would nevertheless be noticeably modified if we took into account the self-positioning of the electorates on a left–right axis. Since the electors of the Social Labor and Democratic Parties in median and northern Europe position themselves clearly much further to the center than do the Mediterranean or French left-wing electors, the spatial distribution of the left would show a stronger peripheral component.

The map shows three categories of areas of strength in which the left has an absolute or at least a relative majority. These can be analyzed not only in relation to the absolute results of the left at the European elections but also in relation to the national averages (map 3—see the Appendices).

• Peripheral areas in which the left-wing stronghold can be explained either by a distrust of the central state and its prevailing structures (e.g., the former GDR, where the PDS appeals to pensioners or intellectuals rather than to workers), sometimes bound to autonomy claims (as in Scotland or in the Pays Basque, where the left is not so strong but holds particularly radical positions) or at least to the maintaining of protection systems (as in Sweden, with a clear general south–north gradient of refusal of European membership for the Scandinavian countries; and in the Norwegian Lapland as well, while in the north of Finland the same mistrust is expressed in agrarian votes). This peripheral left, in which the extreme left is often very specific (maps 4, 5, and 6—see the Appendices), can sometimes rely on political behaviors stemming from agrarian struggles in the regions where the land structures, dominated by the large landowner estates, have slackened the economic development and brought about a sociological secularization as a consequence of the rejection of the church, which is considered too close to the big landowners. This has been the case in the south of the Iberian Peninsula but not in the south of Italy: Here the distrust of the republican legality has led to a clientelist penetration of the state system, mostly through the wheels of the Christian Democracy, as a prolonging of mafia traditions, which from the beginning have expressed refusal of any external authority in economic and political matters. A usual secular positioning on the church-state cleavage (or, in Orthodox countries, the absence of such a cleavage) generally benefits left-wing voting in the regions lacking a marked industrial tradition (the northwestern side of the French Massif Central, Third Italy, Greece). On the whole, places with a high number of farmworkers, share-croppers, forest workers, or sea fishers, having influenced the ideological framing of the region, often sustain left-wing attitudes, even if those social categories are now reduced in the active population distribution.

• Former cores of heavy industry or coal mining—or even, though less specifically, of textile industry—facing an often difficult reconversion process and which we will label "internal peripheries," the more so as most of the time they are deprived of real economic leadership as exists in the Asturias, the Nord-Pas-de-Calais, the basins of Wallonia, the Ruhr, some industrial conurbations in the north of England. The part of the extreme left within the left is significant only when the latter can rely on powerful trade unions, such as in the Nord-Pas-de-Calais. Meanwhile, not all the industrial basins have a left-wing tradition, in particular when their late modernization has involved an early canalizing of the working-class movement by the Catholic church (textile areas of Flanders, Alsace, etc.).

• In addition to the peripheral cores and the redeployment areas, the left is also quite present in most of the central metropolitan areas (sometimes only in a relative measure as compared to the surroundings) (table 11.4). The significance of the left in the urban areas can be explained in different ways. It can be based partly on a prolonging of urban working-class traditions but also today, in other places, on intellectual, radical, and/or postmodern left bases: In almost all metropolitan areas quoted in the table, the left voting is higher than in the national or regional areas of reference. The only exceptions are the Brussels metropolitan region and Stockholm, two capitals with dom-inant tertiary sectors; Lyon; Marseille; and, very markedly, Palermo. The rel-ative weight of the left in these metropolitan regions is frequently induced or at the least supported by extreme left and/or ecologist voting. In the Paris area, the higher-than-average results for the left are due only to the extreme-left and ecologist votes. The periurban wealthy areas are less further to the left than the town centers, where decaying districts, social housing, and neighborhoods on the way to gentrification are concentrated, but also less than their immediate industrial suburbs too (such as the Seine- Saint-Denis and the Val de Marne around Paris). But it is in the wealthy periurban zones that the ecologist vote is most specifically behind the performance of the left.

It appears clearly that the left vote is less and less, if it ever generally was, a class-based vote. On state scale, the gaps between left and right regions are more obvious where they are strengthened by factors of religious observances and strong regional identities, whether based upon language and ethnic identities or upon identities born of economic and rural structures. It is the

Table 11.4 Relative level of left voting and its components in the metropolitan areas relative to a national or regional area of reference (area of reference = 100)

City or metropolitan area	Area of reference	Extreme left	Socialists and Labor	Ecologists	Total left
Vienna	Austria	212	108	162	122
Hamburg	North of West Germany(a)	220	101	157	115
Bremen	North of West Germany(a)	174	119	161	128
Düsseldorf (Stadtkreis)	North of West Germany(a)	138	93	116	98
Cologne (Stadtkreis)	North of West Germany(a)	150	96	183	113
Rhine-Main big cities(b)	North of West Germany(a)	191	81	173	101
Munich (Stadtkreis)	Bavaria	108	100	108	102
Berlin	Former GDR and Berlin	75	110	258	110
Berlin	Germany	277	87	182	130
Basle city	Switzerland	–	167	174	157
Zurich	Switzerland	55	113	82	104
Geneva	Switzerland	835	88	164	151
Brussels-capital et Walloon Brabant	Brussels and Wallonia	69	68	114	88
Randstad Holland(c)	Netherlands	120	123	141	128
Greater London	South of England	199	140	104	134
South-East	South of England	54	78	100	83
Greater London and South-East	South of England	173	129	103	124
Greater London	United Kingdom	90	114	130	115
South-East	United Kingdom	24	64	126	71
Greater London and South-East	United Kingdom	78	105	129	107
Leinster (Dublin)	Ireland	116	140	190	144
Copenhagen metropolitan area	Denmark	127	97	–	104
Stockholm	Sweden	79	79	94	82
Uusimaa (Helsinki)	Finland	82	91	150	111
Oslo	Norway	165	95	221	109
Oslo and Akershus	Norway	129	97	200	104
Paris	France	81	91	161	106
Seine Saint-Denis and Val Marne	France	161	96	118	119
Rest of the Ile-de-France	France	99	94	119	101

Table 11.4 (Continued)

City or metropolita area	Area of reference	Extreme left	Socialists and labor	Ecologists	Total left
Ile-de-France	France	110	94	127	107
Rhône	France	87	94	112	97
Bouches-du-Rhône	France	121	86	93	97
Madrid	Spain	112	100	82	101
Barcelona	Catalonia	109	106	110	107
Barcelona	Spain	44	101	368	104
Seville	South of Spain[e]	130	110	110	113
Seville	Spain	156	128	42	129
Lisbon	Portugal	141	98	160	109
Lisbon and Setúbal	Portugal	166	98	152	114
Big cities of northern Italy	Northern Italy[g]	127	126	119	126
Bologna	Central Italy[h]	38	111	120	98
Florence	Central Italy[h]	140	115	102	119
Rome	Italy	121	110	112	113
Naples	Southern Italy	128	108	175	117
Naples	Italy	113	100	155	106
Palermo	Southern Italy	58	64	100	65
Attica	Greece	137	90	–	102

[a]Former *länder*, except Berlin, Baden-Würtemberg, and Bavaria.
[b]Stadtkreis of Frankfurt, Darmstadt, Offenbach, and Wiesbaden.
[c]Provinces of Noord- and Zuid-Holland and Utrecht.
[d]Copenhagen city and amt Frederiksborg.
[e]Without the Canary Islands.
[f]Turin, Genoa, Milan, Venice.
[g]Without Upper Adige and Emilia-Romagna.
[h]Emilia-Romagna, Tuscany, Umbria, and Marches.

case of Scotland in comparison to England, and of northern England relative to the south. It is also the case in Germany, between the Protestant northwest (where the left is strong as well in the Catholic areas that experienced an early heavy industrialization) and the Catholic south; between the former GDR and former West Germany (but here with the PDS); and, finally, between the north and the south in Spain and Portugal. However, in Portugal a strong Socialist diffusion is now developing toward the north, a stronghold of the right since the revolution; and in Spain, the most industrialized and urbanized areas of the northwest, a traditional stronghold of the right, are being conquered by the regionalist variant of the left in Galicia. As to Italy, even if its hegemonic nature is no longer what it used to be, the left identity of the Third Italy is still alive. The opposition prevails in Austria between the Alpine regions and Vienna, and in Belgium between Flanders and Wallonia.

Among left-wing supporters, the ecologist electors remain very much marked by a specific geography relying on a large intellectual and urban component of progressive inspiration but often stemming from the small bourgeoisie or the Christian tradition historically devoid of a link with the organized Socialist working-class movement (maps 7 and 8—see the Appendices). As mentioned above, the ecologists are well represented in the regions where urban and periurban bourgeois voting is also dominant (as in Belgium, where political ecology has obtained the highest scores in Europe, among the wealthy communes and periurban zones of the southeast of Brussels), but it is also well represented in less urbanized regions with centrist or Conservative voting traditions (the Paris basin and inner west of France, the Belgian southeast, western Austria, Germany's Catholic south, southern Ireland, the south of Finland, etc.). When not on the decrease, the ecologist vote has failed to turn out a success in the south of Europe. The absence of an ecologist party in Denmark (and in Norway) can be explained by the fact that ecologist positions have been adopted by all the parties and that radical political ecology is associated there with the extreme left.

The Right and the Extreme Right

If we add the center and the classical right votes to the extreme right votes, three major areas become visible (maps 9, 14, and 16 see the Appendices).

• A median European Conservative area, whose image would be still reinforced if the cartography took into account the centrist positioning of the left-wing electors within this space, which stretches toward the Rhine and the north of Italy. It is the central, highly urbanized space, corresponding roughly to Roger Brunet's "blue banana." In the Catholic areas, the Christian Democracy has been surviving on two bases (map 11—see the Appendices):

1. state centralization counterbalanced by the historical weights of the local entities, so that the state has not gained a strong hegemony in relation to the church; and
2. an association among the bourgeoisie, the peasant world, and a faction of the working-class movement, especially where this latter has developed and organized itself in the aftermath of *Rerum Novarum,* thus outside the older cores of the industrial revolution, which are denser and more concentrated.

• A more pericentral space in the history and economic development of Europe, some large parts of which have long maintained, or still do, a marked rural character. In these places, the reactionary right claiming an identity withdrawal is dominant (map 17—see the Appendices).[20] Some parts of this space have a strong Conservative tradition, which grows into intense withdrawal positions as a consequence of the opening faced by the local populations (Alpine bloc, Bavaria, and Baden-Würtemberg)—though not always, as in France for instance, where the inward-looking right is now quite present in the Paris basin, in the southwest, and in the southeast, in spite of the centrist republican tradition of those regions, while it is less present in the Catholic inner west. In Italy, the strengthening of the League and of a right that claims a withdrawal into dominant positions incorporates the north of Italy into this zone, even though the latter was not traditionally much positioned to the right except in the northeast.

The regions of Europe where the left is the weakest are thus situated at the crossing of the two preceding spaces, the Catholic south of Germany, the Alpine bloc, and the north of Italy, all areas where the product levels per inhabitant are now the highest in Europe, except for the biggest metropolitan areas, and to which have slipped the center of gravity of the western European economy over the last 40 years.

• Areas dominated by a peripheral Conservative right. These are characterized by the preservation of a significant religious weight (sometimes resulting, in Protestant countries, in antistate attitudes on the part of extremist segments) or by small, more or less marginal rural (or fishing) enterprises. This applies to Finland, Iceland, Ireland, the western coast of Norway, the north of Spain, and, on smaller scales, the western coast of Denmark, the Vendée, etc.

Conclusions

The electoral geography of supranational entities is a perilous exercise, since we are faced with the question of an appropriate classification of the parties, harder than within the national frameworks. Such difficulties are increased with the weakening of the ideological gaps, at least between the catchall parties of the center left and the center right, in a context of dominant neoliberal ideology. Within the big parties of the center right, and sometimes of the center left, the views on some major stakes may diverge between different currents, in particular in relation to European construction and national assertion, sometimes more than between the parties themselves. Even if they have not had the same result as in France, where the right has definitely imploded, a number of Eurocritical stands keep dividing the British Conservative as well as Labour Parties and even the CDU-CSU in Germany. New atypical parties are arising, claiming to defend the interests of the people but on a radical xenophobic right-wing basis. Inversely, a new intellectual and middle-class left is appearing, based partly upon postmodern values, though often showing considerable commitment in the social field.

We have already said why we had deliberately given up mapping the parties according to their belonging to the different groups in the Strasbourg Parliament. In the same way, we have chosen not to map the degree of adherence to the strengthening of European integration. What sense does it make indeed to add British or Nordic abstentionist electors who are supposed to express a rejection of European politics, extreme-left opponents of free-trade and a capitalist Europe, Scandinavian electors wishing to maintain their welfare states, right-wing nationalists, etc.? In our opinion the concept of "tribunician" parties is neither more relevant nor more efficient, especially since not all of those so-called 'tribunician' parties are opposed to the strengthening of European construction—much to the contrary, as some of them demand more social policy.

Europe is said to have lost a part of its pink complexion in the aftermath of the June 13, 1999, elections. We have not studied here the evolution of the electoral scores of the political families relative to the previous European polls. The problems of our classification have been increased, in this case, by the reconstruction of the partisan landscape, in particular in the case of France. Still, the fluctuations of the electoral results from one poll to another do not basically question the relative spatial structures, synthesized in map (see the Appendices) 19, that are at the heart of significance of electoral geography.

Notes

1. In order to provide the most exhaustive view of the European electoral geography in the large sense, that is to say also of that part of Europe that has not lived, after World War 2, in the Soviet sphere of influence (except the former GDR), we have included in our cartography the results of the national elections of the 3 countries that do not belong to the European Union, namely Iceland (for the 1999 legislative elections), Norway (1997) and Switzerland (1999).

2. It could consequently have seemed interesting to map the abstention rates as well as the blank and null votes. However, interpreting such a map would have proved quite hard, considering it would superpose multiple national political traditions, and diverse degrees in the approval or the rejection of the European vision, not to mention some countries like Belgium, where voting is obligatory, even though the non-voters are no longer prosecuted.

3. One has to take account of the gap, sometimes considerable, that can exist between the positioning of the parties as it can be deduced from their electoral platforms, and the positioning of the different parts of their electorates, as it comes out from the Eurobarometers.

4. To the regionalist parties we have added the Swedish Popular Party of Finland, although it is here a question of a-territorial communautarism rather than regionalism.

5. This is a national percentage balanced by the populations (and not by the number of valid votes) of the different countries, including the three countries that are not members of the EU. The same method is used, on the basis of the regional populations, to calculate the percentages in the sub-national entities, such as those appearing in table 2.

6. Outside the EU, Sol has gained one seat, with 0.5 percent of the votes at the latest federal elections in Switzerland.

7. To which is to be added the Norwegian SV (6.0 percent of the votes at the latest legislative elections).

8. So as among the Icelandic Left-Greens (9.1 percent at the legislative elections). The Icelandic socialists and leftist feminists are grouped, together with the social democracy, in a Union (26.8 percent) that we have classified in the classical socialist left, although the Icelandic communists, having early escaped the Soviet influence, have played an essential role in its shaping.

9. Outside the EU, the Socialist family groups the Norwegian Labor Party (35.0 percent), the Swiss Socialist Party (22.5 percent), and Iceland's left-wing Union.

10. To this must be added, outside the EU, the Swiss ecologist party (5.0 percent).

11. As well as, outside the EU, the Left-Greens (9.1 percent) in Iceland.

12. This reserve allows us to include in this group the British Conservatives, many of whom are Euroskeptical. And even if they are explicitly anti-European, some other parties have to be included in this group as well, because of their other characteristics. Two of these parties have deputies in Strasbourg: JuniBevaegelsen mod Union in Denmark (16.7 percent of the national votes) and the UK Independence Party in Britain (6.5 percent). It is significant that they do not belong to the two big center and center right groups of this assembly, the EPP and the EDRL, but to the EDD (Europe of Democracies and Diversities).

13. Herein are included the countries divided between Catholics and Protestants: Germany, the Netherlands, and, outside the EU, Switzerland, although as much the CDU-CSU as the Dutch CDA or the Swiss PDC enjoy stronger positions in the Catholic parts of those countries.

14. One can argue about the genuine Christian Democratic character of the main party included in this trend, namely the German CDU-CSU, as well as the Austrian ÖVP, as they could be considered most of all as big catchall parties of the center right, lacking a specific trade unionist wing. We have nevertheless decided to include them in the Christian Democratic family. The CDU-CSU has a partial affiliation with the German pre-Nazi Zentrum; It is overrepresented in the German Catholic areas; its electoral platform keeps a reference to Christian values; and it is bound to the working class through specific associations. The ÖVP is situated in the historical continuation of the Austrian Social Christian Party founded by Karl Lüger in 1889 and one of its constitution league represents in fact the Christian wing of the Austrian unitarian trade

union, the ÖGB. Both the CDU-CSU and the ÖVP play a significant part in the European Union of the Christian Democratic parties.

15. Two Spanish regionalist parties, the Catalan UCD and the Basque PNV, could be included in the Christian Democratic family. The UCD representative belongs indeed to the EPP group in Strasbourg, but the results for this party cannot be dissociated from those for the CDC, with which the UCD formed the common CiU list at the European elections. As to the obviously cross-class PNV, it has joined the Greens/European Free Alliance.

16. To which one can add the Popular Evangelic Party in Switzerland (1.8 percent).

17. In the proper sense of the word, that is to say through promoting a return to more traditional values supposed to represent a mystified past.

18. The political sociologies of the electorates of the Austrian FPÖ and the Swiss UDC/SVP are probably very much alike.

19. As well as the UDF in Switzerland (1.3 percent).

20. Such withdrawal positions are found in other pericentral zones, though expressed differently: To the votes for the DF in Denmark can be added other political expressions of the refusal of Europea adhesion in this country. In Ireland, the identity vote is dominant, even if it is integrated by the traditional political parties (in this case devoid of anti-European complexion), insofar as an increased European weight can but counterbalance the British one. In the same way, their distrust of London leads Scotland and Wales to a more pro-European vote than the English one.

References

Hermet G., Gottinger J. T., and Seiler D. L. eds. (1998), *Les partis politiques en Europe de l'ouest*. Paris: Economica.

Lipset S. M., and Rokkan S. (1967), *Party Systems and Voters Alignments*. New York and London: Free Press, Collier, Macmillan.

Perrineau P., Ysmal C. et al. (1995), *Le vote des Douze*. Paris: Département d'Etudes Politiques du Figaro et Presses de Sciences Po.

Rokkan, S. (1970), *Citizens. Elections. Parties*. Oslo: Universitetforlaget.

Seiler, D. L. (1980), *Partis et familles politiques*. Paris: PUF.

Seiler, D. L. (1982), *Les partis politiques en Europe*. Paris: PUF.

Todd, E. (1951), *L'invention de l'Europe*. Paris: Seuil.

Vandermotten C., Colard A., and Delwit P. (1996), "Géographie électorale de la gauche en Europe des années quatre-vingt et quatre-vingt-dix," in Lazar M. (ed.), *La gauche en Europe depuis 1945. Invariants et mutations du socialisme européen*. Paris: PUF.

Van Laer, J. (1984), *200 millions de voix*. Bruxelles: Société Royale Belge de Géographie.

CHAPTER 12

Electoral Participation and the European Poll: A Limited Legitimacy

Pascal Delwit

Introduction

The poll of July 13, 1999, reproduced what various mass-media, political, and sociologist environment observers had forecasted, without any risk of failure: An important abstention was registered at the European elections. The "Party of the Abstention" preceded all other parties at the European level, both in terms of the national and the supranational discussion. We can even affirm that a weak electoral participation is the common point of this European poll. This election revealed certain common tendencies, but the most visible and important one was, without any doubt, the absence of an electoral mobilization of the European Union citizens.

Certainly we can point out, like Richard Corbett, Francis Jacobs, and Michael Shackleton, that the electoral participation in the European polls is perceptibly superior in comparison with the United States national polls—even the presidential poll.[1] Nevertheless, in comparison with the turnout in different member states of the European Union, the political participation of the European citizens is obviously weaker. On the other hand, the political dynamic of democracy is different in Europe than it is in the United States.[2]

In this direction, first we are going to focus our analysis on the substance of this weak electoral participation for this fifth European poll based on universal suffrage and then we are going to suggest several elements in order to conceive this abstention from the perspective of the classical literature regarding political participation—with the main accent on electoral participation. Finally, we will focus on the consequences of this nonmobilization for the European deputies delegation.

A Falling Participation

From the first European elections, the electoral participation in the nine member states was under the average registered during the national elections. The average of the abstention touched 37.57 percent, with the peak in the United Kingdom (67.23), in Denmark (52.16), and in the Netherlands (41.92). Despite an EC promotional campaign to increase electoral participation, the electoral mobilization was not in top form for the first European elections to the European Assembly on the basis of universal suffrage. The following polls confirmed this lack of interest. In 1984, the abstention rate grew to 40.83 percent, or 39.32 percent, if we take into account the Portuguese and Spanish performances. Five years later, the lack of interest was even more visible: 43.83 percent of European citizens did not express their votes despite the adoption of the Single European Act, which was supposed to bring a new dimension to the European Community. The 1994 poll marks the only stabilization in this area—the abstention rate was similar to that registered in 1989: 43.22 percent, or 43.34 percent if we take into account the three European polls from Finland, Sweden, and Austria during this legislation.

The European election on July 13, 1999, registered a new low in electoral participation. For the first time, the abstention rate passed over 50 percent; an average of at least one voter out of every two showed a lack of interest in representation at the European Parliament level.

Table 12.1 Turnout at five European polls

	1979	1984	1989	1994	1999
United Kingdom	32.77	32.92	36.92	36.49	24.02
Germany	65.70	56.80	62.30	60.02	45.19
Austria				67.73	49.40
Denmark	47.84	52.38	46.15	52.92	50.46
Spain		68.53	54.72	59.14	64.38
Finland				57.59	30.00
Ireland	63.61	47.56	68.28	43.98	50.70
Italy	85.69	82.90	81.60	74.77	69.76
Netherlands	58.08	50.91	47.53	35.69	29.95
Portugal		70.38	49.70	35.67	40.03
Sweden				41.62	38.84
France	60.74	56.82	48.79	52.71	46.76
Belgium	91.29	92.19	90.73	90.56	90.96
Greece		80.54	79.97	71.24	70.27
Luxembourg	88.90	88.80	87.60	88.54	86.63
EC-EU (9)	62.43				
EC-EU (10)		59.17			
EC-EU (12)		60.68	56.17	56.78	
EC-EU (15)				56.66	49.62

Analyzing these data, we can observe an important decline in the turnout. In 1979, 22 percent of the European states registered a turnout of only up to 50 percent; in 1989, 41.67 percent of the European states account for the same turnout; and in 1999, 53.33 percent. Concerning the countries registering an abstention rate superior to 50 percent, it is important to underline the presence of the three biggest member states: the United Kingdom (75.98 percent of abstention), France (54.24), and Germany (54.81).

Analyzing the data from the basis of the pyramid, it is important to stress that the number of countries where the turnout was superior to 70 percent is the same in 1999 as in 1979—taking into account also the enlargement of the EU with six new countries.

Still, we have to moderate these "good rates." Belgium, Greece, and Luxembourg are small or medium-size member states and the vote is compulsory and systematically joint with the national elections (e.g., Luxembourg). In Italy, the vote is quasi-obligatory. Taking into account these nuances, the comparison with other national situations is not fully justified. Due to the link between the turnout in the EU elections and the compulsory vote, the context should be analyzed from a different perspective.

The turnout for the 12 states where the vote is not compulsory touches only 48 percent, while it registers 79.36 percent at the level of the European member states where voting is obligatory.

Making a comparison between the turnouts at the level of the member states' national elections and those registered at the European elections level, an important gap makes itself visible. On average, this is a gap of about 27.44 points. Beyond what we have already mentioned for the *big* countries, it is

Table 12.2 Number of states in conformity with the participation slices at the European level

	1979	1984	1984 plus the new member states	1989	1994	1994 plus the new member states	1999
0–30%	0	0	0	0	0	0	3
31–50%	2	2	2	5	4	5	5
51–70%	4	4	5	3	4	6	4
71–90%	2	3	4	3	3	3	2
More than 90%	1	1	1	1	1	1	1
N	9	10	12	12	12	15	15

Table 12.3 Abstention rate on the basis of the existence or nonexistence of compulsory voting

	1979	1984	1989	1994	1999
EC-EU (noncompulsory voting)	61.29	58.99	54.19	55.19	48.00
EC-EU (compulsory voting)	91.23	86.21	84.94	79.71	79.36

Table 12.4 Difference between the latest turnout at the national level and the turnout at the European polls of July 13, 1999

United Kingdom	47.46
Germany	37.01
Austria	36.58
Denmark	35.48
Spain	13.00
Finland	35.27
Ireland	17.70
Italy	13.01
Netherlands	43.40
Portugal	26.26
Sweden	42.55
France	21.20
Belgium	−0.38
Greece	6.07
Luxembourg	−0.12
European Union (15)	27.44

symbolically important to observe that this gap is especially visible for the three most recent member states. In each case, the figures are clearly superior to the European average: Austria (36.58), Finland (35.27), and Sweden (42.55).

The abstention was very important on July 13, 1999. It touched a higher point. Without any intention to plead for a catastrophe interpretation, it is still important to emphasize that these scores are even bigger than stressed in the table above. The abstention rate is calculated on the basis of registered voters. In several countries, the registration is automatically done. In other member states, like France, the registration is a voluntary process of the individual voter. Some authors evaluate the number of nonregistered voters at more or less 10 percent of the potential voters. Or, as observe Pierre Bréchon and Bruno Cautrès in their analysis of the Grenoble example, "the issue regarding the interpretation of the non-registration is as important as the interpretation of the abstention."[3] The two authors have examined how much elements of socialization and politicization influence the matter of registration or nonregistration. Alain Lancelot, in his classic study on the abstention, explains nonregistration through five behaviors: ignorance, laziness, indifference, evasion, and refusal.[4] In their analysis of the abstention, Françoise Subileau and Marie-France Toinet stress the lack of social integration: "Concerning the nonregistration matter, the characteristics of a precarious professional status play a relatively important role."[5] Nevertheless, as proved by several studies, the extent of nonregistration/abstention should be well balanced by the existence of fake registered voters,[6] especially in the urban areas.[7]

Table 12.5 Turnout calculated on the basis of the valid votes

	1979	1984	1989	1994	1999
United Kingdom	32.67	32.91	36.81	36.49	24.00
Germany	65.30	55.90	61.60	58.56	44.51
Austria				65.41	47.91
Denmark	46.84	51.78	45.60	52.07	49.13
Spain		67.03	53.52	58.19	62.78
Finland				55.02	29.86
Ireland	61.18	46.42	66.55	43.22	49.06
Italy	82.92	79.11	75.73	69.33	62.82
Netherlands	57.78	50.51	47.23	35.58	29.90
Portugal		70.19	49.54	34.53	38.69
Sweden				40.95	34.53
France	57.53	54.71	47.39	49.94	43.99
Belgium	80.04	82.09	83.13	82.74	84.74
Greece		80.54	79.97	71.24	70.27
Luxembourg	80.30	80.60	78.60	79.74	79.13
EC-EU (9)	60.30				
EC-EU (10)		58.94			
EC-EU (12)		57.34	56.66	54.67	
EC-EU (15)				54.58	47.43

We should also take a glance at the total of "white" bulletins and null ones; they are not expressing a vote for one of the political forces in competition. These votes can be ranged among the abstentions, sometimes qualified as "civic abstentions."[8] Concerning the countries with compulsory voting, Alain Lancelot evokes the white and null bulletins as expressions of a real abstention. In fact, the number of white and null bulletins is particularly high in the three countries with compulsory voting.

If we calculate the turnout on the basis of the valid votes and not on the basis of the expressed votes, the scores are even weaker—a loss of 2 percent is registered, and the electoral turnout falls from 49.62 percent to 47.43 percent.

The Reasons for Nonparticipation

A Drop in Electoral Participation

A first nuance has to be taken into account concerning the analysis of the abstention at the European level. A general view on this topic proves that the abstention rate registers progress during the last quarter of the century. The European elections follow this trend, which is visible at both the national level and the regional one. Jean Chiche and Elisabeth Dupoirier evoke the "Abstention Party" in the French national elections in June 1997: "For the three latest parliamentary elections of this decade, the voters registered on the electoral lists who did not present themselves at the polls created 'the first party of France.'"[9] Françoise Subileau points to the new crisis situation: "Comparing similar elections brings us to a striking conclusion: the abstention increased significantly during the last twenty years"[10]—nevertheless her statement has been moderated by other political scientists judging the electoral participation from a larger historical perspective.[11]

Twelve of the 15 member states of the European Union share a linear decline: a more significant [pronounced] abstention rate for the last election of the 1980s than for the last election of the 1970s, and an even weaker participation for the last elections than for the last poll of the 1980s. Denmark is an almost perfect example of the status quo. The evolution of the electoral participation in Spain is related to the specificity of the democratic transition. From this perspective, the particularly weak abstention of the 1979 national elections has to be carefully analyzed. Even if the latter is considered apart from the abstention rate registered at the European level, it is still important to underline that for the last 20 years, they have shared a configuration based on an electoral participation in decline.

The Abstention and Institutional Constraint

The studies dedicated to the abstention generally mention institutional constraints, such as the electoral system. Since 1979, several authors have explained the weak participation at the European-elections level as the consequence of the lack of harmonization of the electoral system among the different member states. A harmonization in this field has been perceived as a necessity from the beginnings.[12]

It is obvious that the existence of different electoral systems makes difficult the construction of a supranational partisan building.[13] This heterogeneity is conceived not only as a limitation on the powers of the European Parliament but also as a paralyzing factor in relation to the coherence and structure of the European political groups. An important modification of the British electoral system could profoundly influence the ideological composition of the European Parliament and the balance of the inner structure of each group. The argument of the electoral systems' heterogeneity is no longer used to explain the participation issue. A certain convergence of the electoral systems is operating—the adoption of the proportional system at either the national constituency level or the regional constituency level is today a fact. "The last exception" is no more. The United Kingdom introduced a profound modification of its electoral system for the European elections, adopting a proportional system on the basis of regional constituencies. Nevertheless, this

Table 12.6 The tendency toward decline in the electoral turnout in Europe

	Last poll of the 1970s	Last poll of the 1980s	Last poll of the present	Distance between the last poll of the present and the last poll of the 1970s
United Kingdom	(1979) 76.00	(1987) 75.42	(1997) 71.48	−4.52
Germany	(1976) 90.74	(1987) 84.33	(1998) 82.20	−8.54
Austria	(1979) 92.24	(1986) 90.46	(1995) 85.98	−6.22
Denmark	(1979) 85.62	(1988) 85.2	(1998) 85.94	0.32
Spain	(1979) 68.13	(1989) 70.80	(1996) 77.38	9.25
Finland	(1979) 75.31	(1987) 72.06	(1999) 65.27	−10.04
Ireland	(1977) 76.32	(1989) 68.51	(1997) 68.01	−8.31
Italy	(1979) 94.28	(1987) 90.49	(1996) 82.77	−11.51
Netherlands	(1977) 88.00	(1986) 85.76	(1998) 73.35	−14.65
Portugal	(1979) 87.54	(1987) 70.38	(1995) 66.29	−21.25
Sweden	(1979) 90.72	(1988) 85.96	(1998) 81.39	−9.33
France	(1978) 83.24	(1988) 66.18	(1997) 67.96	−15.28
Belgium	(1978) 94.87	(1987) 93.37	(1999) 90.58	−4.29
Greece	(1977) 81.11	(1979) 79.59	(1996) 76.34	−4.77
Luxembourg	(1979) 88.85	(1989) 87.39	(1999) 86.51	−2.34

transformation did not influence an augmentation of the turnout for the 1999 European elections. Only 24.02 percent of United Kingdom citizens expressed themselves, in comparison with 36.49 percent in 1994.

The importance of the abstention rate challenges the link established by certain authors between electoral systems and electoral participation. Through a successful management of the useless votes, the diversity of the electoral offers should determine a superior mobilization in comparison with the elections on a majority basis or in a mixed electoral system.[14] New political groups or organizations could fight on the electoral background and (re)assemble the citizens sheltered under the abstention cover due to the lack of attraction of the old parties.[15] André Blais has established, in two studies dedicated to the electoral abstention, a statistical correlation between the proportional system and electoral participation.[16]

In reality, as Arend Lijphart has already pointed out in a very sober way, this correlation was not verified at the European elections level. Although all the 15 member states of the European Union adopted a proportional system for the European elections, participation touched its weakest rate. We can even ask ourselves whether the absence of polarization of the European electoral system canceled a certain effect of the proportional system at the national elections level. Some authors have written about an *effet pervers* of the list system at the

European level. This is also the point of view of Jean-Louis Quermonne concerning the French case: "This system, criticized once in the decision of the Constitutional Council of December 29 and 30, 1976, changed the whole poll issue—it moved the deputies away from the voters and dispersed them among the parliamentary groups of the Strasbourg Assembly. This situation is going to diminish their influence in comparison with the influence of their colleagues from other countries taking part in the Party of European Socialists or the European People's Party."[17] The French electoral system is not the only reason for the dispersal of political forces in different groups. In Germany, the same electoral system generates different effects. Still, the debate goes on.

Beyond the issue of the electoral system in the member states, simple institutional facts can influence the expression of the abstention. An analysis of the weak participation at the 1994 European elections has underlined that there was a particularly strong abstention in the countries where the voting sessions were organized during the week, in comparison with those organized during Sunday. In addition, June is a month of tourism for an important part of the nonmarginal society.[18] We can add that June is also a period of exams, less favorable to the mobilization of young voters still in attendance at school. At the same time, there was no national poll, as had been the situation in the past, that could have "stimulated" the turnout in the countries without compulsory voting.[19] Nevertheless, this argumentation is not the essential one.

Less Obvious Issues

The "meaning" that the European elections possessed can be an explanation of the lack of participation at the European level. Debating the logic of abstention in 1968, Alain Lancelot wrote: "If we are considering the electoral behavior as the output of the subject (the voter) to a question (the consultation), we can suppose that the answer of this output (eventually the abstention) depends on the inner meaning of the question put and on the way in which it is formulated. The analysis confirms that the panel of options available for the elections, the acuteness of the competition, and the extent of the consultation have a certain influence on the abstention volume."[20] Over thirty years later, these patterns of analysis allow us to understand the lack of citizens' involvement in European elections and the strategic abstention that imposes itself.

Schematically, we can reduce the issue of the European poll to two important questions that influence the voter's behavior: What kind of future will exist for the European Union? and What kind of majorities will be needed for what kind of policy or policies?

Basically, the voter does not possess the ability to answer these questions. In addressing the first question, we might ask, Are there really different visions of the "common European good"? The answer does not seem obvious. With the exception of the anti-European forces—several extreme right parties and nationalist groups—which emphasize speech-oriented against the supranational perspective, there is not a very clear possibility of choice among different options regarding the European Union. The programs are very general and not very distinctive at a national level. The differences are located mainly among the supranational families, but they are not very visible at a national level and *a fortiori* for the voters.[21] Although the European question often divides the parties, there is no clear position taken by them—especially regarding the institutional chapter—in order to diminish the internal disputes. The competitive characteristic of the European polls is reduced to a panel of generalities, not concrete enough for the citizens.

Nevertheless, in regard to the second question, every parliamentary election is connected to the choice of deputies and to the relation with the government that the voter wants to promote or sanction. This dimension is absent at the level of the European polls. The European Parliament has very little, if anything, to do with the image of national parliaments.[22] In fact, they have quite different forms of parliamentarianism. The European Assembly does not sanction a government positively or negatively. There is no easily identifiable opposition or majority inside this assembly. Members of the Parliament are well aware of this fact. Their attitude differs from that of "pressing a button," as they have the opportunity to have "high level" debates and deliberations. But is this introspective vision necessary and comprehensive for the European citizens? Why and for what are we voting? Without a majority in government and/or in the Parliament, there is no easy answer to our question. This is not only due to the absence of a "majority," but also because the question is related to different political, economic, cultural, and educational choices. In different terms, the question should be, What kind of policy are we voting for? No single answer belongs to everybody. The influence of the European Parliament in the decision-making process is not very important—despite the increase in its prerogatives—and the partisan reaction, when there is one, is quite evasive. So, without knowing either for what or for whom he is voting, the voter does not go to the polling station.

Could this mean that the European question is less politicized?[23] Not at all. On the contrary, it is more and more politicized, but this dimension is not seen at the European elections. Authors have underlined the specific nature of the European poll since 1979, that is, there is no European election, but a panel of national elections that can be characterized as "second-order" elections,[24] "intermediary" elections,[25] or "midterm" elections.[26] In corroboration, the elections of 1984, 1989, and 1994 were primarily panels of national elections and secondly European elections. Despite the efforts of the

European federations of parties to establish electoral manifestos that were more substantial and more focused on the European matter, "Europe," and "the European Union," the challenge of European construction continued to be only secondary or even absent issues in the campaigns and polls. From this point of view, the elections of June 13, 1999, fully merit the label of "intermediary," having been first of all an evaluation of internal political forces. If the European poll can provoke effects, those effects are still far away. Michel Rocard proved it in 1994, and the conflicts connected to the results of this European election in Germany and Italy certify it. In other countries, the situation has a minimal impact, which contributes to a lack of mobilization.[27] It is clear that in the Scandinavian countries, the Social Democrats register very weak scores due to the difficult relationship with their social base regarding the European issue. The lack of interest of the British voters in the European elections is also very important. Due to the lack of participation, it is difficult to extrapolate these results to the level of national political life.

But there are different forms of mobilization and/or participation. Since the signing of the Maastricht Treaty, the referendums concerning European issues have mobilized a larger part of the society. The two Danish referendums of April 1992 and April 1993 and the French referendum of September 1992 proved that there was a strong politicization of the debates and that there was a real issue related to the consultation: The interest and the participation of the people were there.[28]

By the same logic, there has been a multiplication of social-expression movements related to the issues of European construction. Several authors have described the strong French social movement of December 1995 as a reaction generated by European construction and as "the first revolution against globalization,"[29] while, essentially, the mobilization of support for the French strikers "reached its peak in those places where in 1992 the 'No' votes were the majority at the time of the referendum on Maastricht."[30] Moreover, this social situation influenced the French parties. For several weeks afterward, the European question divided the Socialist Party. On January 18, 1996, Philippe Séguin, president of the National Assembly, characterized the treaty of Maastricht as a "historical stupidity."[31]

A Noncampaign

The studies regarding electoral participation emphasize the importance of context. In his pioneering researches, André Siegfried established the difference between polarization and stakes, which could characterize the polls as "calming polls" or "combat polls."[32] Since the first researches on the electoral phenomenon, the salience of the stakes and polarization of the campaigns have influenced the turnout. We have already pointed out the stakes issue. For the

European polls, the lack of visibility concerning the stakes was reinforced by the feeling offered by the main actors of the political scene. There was a lack of mobilization due to the political will of the parties. There was no real will to promote a genuine political campaign either at the level of the internal structures of the party or at some external level. Symbolic and material involvement rests a simple expression away—sometimes we have difficulty accepting this process as an election. An important part the European citizens ignored, till the election moment, was that a European poll was going to be organized.

This weak involvement of the political groups determines a lack of involvement by the other protagonists of the debate, especially among the media—this bidirectional process is not only a receiver of feedback but is also a cause. The interest of the press in an event is proportional to the interest manifested by the national parties, and vice versa. Concerning the European elections, the involvement effect was low on the part of both, which increased a feeling of nonimportance regarding the poll and a symbolic underappreciation of the need, the duty, or the imperative of voting.

Consequences

A Weak Legitimacy

The link between universal suffrage and the legitimacy process is a classic one. Despite elitist theories, there is also a relation between the extent and the nature of electoral participation and the legitimacy of the Assembly or of the elected.[33] One of the arguments of the partisans of universal suffrage for the elections to the European Parliament was precisely focused on the new legitimacy of the Assembly. This increased legitimacy was perceived as a bonus in the management of the institutional relations with the Council of Ministers, the European Commission, and later the European Council. The European Assembly elected through a universal suffrage should have underlined its difference in comparison with the other international organizations formed indirectly by co-optation of the national deputies.[34]

Soon after the European elections of 1979, different authors stressed a more or less important failure of the legitimization of the Assembly through direct suffrage. Nevertheless, the explanation advanced in order to justify this failure was built on the European Parliament's lack of powers. It has been advanced that an augmentation of the Parliament's competences should diminish the abstention rate: "As long as the EP is not given more powers, i.e., as long as very little, if anything at all, is at stake for the political decisions of the EC system at European elections, turnout at European elections will remain below the turnout of other national second-order elections, contributing only in an ambiguous way to the legitimacy of the EC."[35] Is this argument still valid? Certainly less so. During the last two administrations, the European Assembly acquired real and important prerogatives related to the decision-making process of the EU. Despite this new situation, the abstention rate is even higher than before.

"Un malaise démocratique"

An analysis of the European Union offers us a paradoxical picture. The influence and importance of EU policies continually influence the political life of the member states and their internal regulations. At the same time, citizens are continuing to be less involved in the EU. This situation is unhealthy for a democracy. Certainly, we can partially agree that the abstention during the electoral consultation cannot be seen as a sign of depoliticization.[36] But the problem is still there. There is a real "malaise démocratique,"[37] an unhealthy lack of interest in the European issue proved by numerous citizens, at both the national and the European level.

This phenomenon is more important in connection with a certain inequality of representation. Certainly, all the social and cultural fields of society participated in the phenomenon of strategical abstention—due to the salience of the stakes, the nature of the poll, the conditions of the electoral offer, or institutional matters. But we have known for a long time that abstention is connected to social influences, as "status abstention."[38] The socially privileged categories are always more involved than the others, as the figures of the Sofres-Cevipof poll for the 1997 French elections prove. Proletarian workers and the jobless are the least involved.

Of course, social background can be the only basis for argument, but it is a strong one for political nonparticipation, suggesting (in the French case) that a "link between social crisis and abstention exists in certain sectors of the population where the extent of the social crisis influences its translation into electoral terms. The mobilizing effects of the polls become powerless and cannot correct the expression of a feeling of social marginalization."[39]

The implications of this observation are anything but neutral for policies or the electoral final figures. For instance, a certain relationship has been proposed between the importance of the abstention and the figures of the left or center-left parties[40]—the weaker the political participation, the more the figures of these political families are important. For the European elections, this relation should be carefully analyzed. Other studies have shown a correlation between the electoral force of the postmaterialists and the weakness of the abstention.[41] Or the European polls can prove results in the opposite direction. The Green Parties have always scored high at the European level despite a weak electoral participation. Nevertheless there is a real question about the representation of the economically and/or politically disadvantaged and the political forces that

Table 12.7 Electoral abstention and social attributes

	1986	1993 (first tour)	1997 (second tour)	Difference 1986–1997
Total	21.5	31	31.5	+10
Sex				
Male	19	30	30	+11
Female	24	31	33	+9
Age				
18–24 years	25	40	40	+15
25–34 years	27	38	40	+13
35–49 years	21	25	30	+9
50–64 years	17	26	23	+5
65 years and more	19	29	28	+9
Profession				
Agricultural worker	12	14	27	+15
Small commerce, artisan, small society manager	17	22	37	+2
Senior executive managers, superior intellectual profession	21	37	23	+2
Intermediary Profession	19	27	30	+11
White-collar worker	27	32	31	+4
Blue-collar worker	24	36	38	+14
Inactive, retired	19	31	29	+10
Activity field				
Self-employed	16	20	32	+16
Public services	20	27	30	+10
Private services	26	36	38	+14
Jobless	27	34	40	+13
Votes for the second tour of the presidential election—before the election				
Left candidate	19	29	24	+5
Right candidate	11	21	26	+15
Abstention, white, null	40	47	59	+19

are closed to their positions. Generally speaking, this question is so often put that, to counter the effects of the lack of the participation, Arend Lijphart has recommended the use of compulsory voting in order to answer this democratic dilemma: "Compulsory voting cannot solve the entire conflict between the ideals of participation and equality, but by making voting participation as equal as possible, it is a valuable partial solution."[42]

Conclusions

The 1999 European elections registered a new drop in the turnout—one voter for every two did not go to the polls. The weak participation in the European elections is not new but its numbers are increasing.

This nonmobilization can be explained partially through the material and institutional conditions of the poll organization—disadvantageous timing, a questionable electoral system, a limited number of deputies to elect, etc. These new scores have to be reintegrated into the larger table of the falling scores of participation at the national level over the last 20 years.

But the nature of the problem is different. The relation between the European Union and the citizen is in question. Despite a progressive augmentation of the European Parliament's prerogatives, participation is falling. The politicization of the European polls is a specific one. The actors of the political and media environment do not identify these elections as having any specific stakes, other than choices in European matters or, globally speaking, the evolution of European construction. As we have already seen, the problems are not very well researched and differentiated. Essentially, the European elections remain a national poll of "evaluation," their importance dependent upon the case in each individual state. Their effects are searched for only on this level—the reinforced position of the European People's Party in comparison with the Party of European Socialist met with a lax reception by the political elite. Taking into account this whole landscape, it is not surprising that citizens do not go to express their votes—all these factors contribute to weak electoral mobilization.

This situation is not without danger. It is the sign of both a profound paradox and a dysfunction. The impact of the choices registered at the European Union level reaches its peak in relation to the different national policies, but the abstention rate at the European elections also increases. Certainly, it is obvious that the European Parliament is only a part of the democratic control of the decision-making process at the European level, but it is a symbolic and materially important one. The legitimacy of the European Assembly is affected as well as the legitimacy of the European decision-making process.

Notes

1. Richard Corbett, Francis Jacobs, Michael Shackleton, *The European Parliament*, Catermill, 1995 (third edition), p. 30.
2. In reality, the large abstention in the United States is due to "the political strategy in order to discourage participation" consider Françoise Subileau et Marie-France Toinet in their comparative study between the abstention in France and in the

United States. Françoise Subileau, Marie-France Toinet, *Les Chemins de l'abstention. Une comparaison franco-américaine*, Editions de la Découverte, 1993, p. 64.

3. Pierre Bréchon, Bruno Cautrès, "L'inscription électorale : indicateur de socialisation ou de politisation," *Revue française de science politique*, Août 1987, no. 4, vol. 37, p. 505.

4. Alain Lancelot, *L'abstentionnisme électoral en France*, Presses de la Foundation Nationale des Sciences Politiques, 1968, p. 34.

5. Françoise Subileau, Marie-France Toinet, *op. cit.*, p. 114.

6. Alain Lancelot, *op. cit.*, pp. 42–43.

7. Annick Percheron, "Peut-on parler d'un incivisme des jeunes ? Le cas de la France," *International Political Science Review*, 1987, n. 3, p. 275.

8. This expression belongs to Françoise Subileau in *Le Monde*, May 16, 1997.

9. Jean Chiche, Elisabeth Dupoirier, "L'absention aux élections législatives de 1997," in Pascal Perrineau, Colette Ysmal, *Le vote surprise. Les élections Législatives des 25 mai et 1er juin 1997*, Presses de Sciences Po, 1998, p. 141.

10. *Le Monde*, May 6, 1997.

11. Reacting to François Subileau, Pierre Bréchon put the present abstention in the new context: "1986 corresponds to a 'combat election.' Taking into account the evolution of the abstention rate over a long period of time ... we can observe that certainly the abstention rate today is high but it is not exceptional in comparison with the past." Pierre Bréchon, "Les inconnus de l'abstention," *Le Monde*, May 21, 1997.

12. *Traités instituant de les Communautés européennes. Traités portant révision de ces traités, Actes relatifs à l'adhésion*, Brusells: Office des publications officielles des Communautés européennes, 1978, p. 333.

13. John Smith, "How European are European elections?" in John Gaffney, *Political Parties and the European Union*, Routledge, 1996, p. 276.

14. Richard Rose, "Evaluating Election Turnout," *in Voter Turnout from 1945 to 1997. A Global Report on Political Participation*, Institute for Democracy and International Assistance, 1997 (second edition), p. 38.

15. Markus M. L. Crepaz, "The Impact of Party Polarization and Postmaterialism on Voter Turnout," *European Journal of Political Research*, 1990, no. 2, vol. 18, p. 193.

16. André Blais, R.K. Carty, "Does Proportional Representation Foster Voter Turnout?" *European Journal of Political Research*, 1990, no. 2, vol. 18, pp. 167–181; and André Blais, Agnieszka Dobrzynska, "Turnout in Electoral Democracies," *European Journal of Political Research*, 1998, no. 2, vol. 33, pp. 239–261.

17. Jean-Louis Quermonne, "L'adaptation de l'Etat à' l'intégration européenne," *Revue du droit public et de la science politique en France et à l'étranger*, in the special number, *Les quarante ans de la cinquième République*, 1998, no. 5–6, p. 1410.

18. Jean Blondel, Richard Sinnott, Palle Svensson, "Representation and Voter Participation," *European Journal of Political Research*, no. 2, vol. 32, 1997, pp. 243–272.

19. In the Luxembourg, the European elections are organized at the same time as the national ones. Without this exception, only Belgium has this simultaneity of

European and national elections. Taking into account that in Belgium there is compulsory voting, the impact on participation is less important.

20. Alain Lancelot, *op. cit.*, p. 95.

21. See, for example, Pascal Delwit, "Les divergences de visions partisanes sur le rôle du Parlement européen," in Pascal Delwit, Jean-Michel De Waele, Paul Magnette (eds.), *A quoi sert le Parlement européen? Stratégies et pouvoirs d'une Assemblée transnationale,* Complexe, 1999.

22. Pascal Delwit, Jean-Michel De Waele, Paul Magnette, "Vers un nouveau modèle de parlementarisme?" in Delwit, De Waele, Magnette (eds.), *op. cit.*

23. Patrick Lecomte, Bernard Denni, *Sociologie du politique,* Grenoble: Presses Universitaires de Grenoble, 1990, p. 14.

24. Karlheinz Reif, "Ten Second-Order National Elections," in Reif, *Ten European Elections,* Gower, 1980, pp. 1–36.

25. Jean Charlot, *La politique en France,* Le livre de poche-référence, 1994, p. 151.

26. Neill Nugent, *The Government and Politics of the European community,* Macmillan, 1993 (second edition), p. 145.

27. See on this topic the arguments of Marcel van Egmond, Nan Dirk de Graaf, and Cees van der Eijk regarding the Dutch elections for the last quarter century. Van Egmond, de Graaf, and Eijk, "Electoral participation in the Netherlands : Individual and Contextual Influences," *European Journal of Political Research,* 1998, no. 2, vol. 34, pp. 281–300.

28. For example, the turnout at the referendum of September 1992 was 70.51 percent, compared with 55 percent at the European elections of 1994.

29. *Le Monde,* December 7, 1995.

30. Pascal Perrineau, Michel Wieviorka, "De la nature du mouvement social," *Le Monde,* December 20, 1995.

31. *Le Monde,* December 18, 1996.

32. André Siegfried, *Tableau politique de la France de l'Ouest sous la Troisième République,* Armand Colin, 1964 (2ᵉ édition).

33. Françoise Subileau, "Une participation en baisse depuis dix ans," *Revue politique et parlementaire,* mars-avril 1998, p. 54.

34. Richard Corbett, Francis Jacobs, Michael Shacleton, *op. cit.*, p. 13.

35. Karlheinz Reif, "National Electoral Cycles and European Elections, 1979 and 1984," *Electoral Studies,* December 1984, no, 3, vol. 3, p. 253.

36. Jacques Ion, *La fin des militants?* Editions de l'Atelier, 1997, p. 103.

37. Pascal Perrineau in *Le Monde,* July 1, 1999.

38. Dominique Andolfatto, "Quand les abstentionnistes s'expriment," *Revue politique et parlementaire,* July-August 1992, p.41.

39. Jean Chiche, Elisabeth Dupoirier, *op. cit.*, p. 149.

40. Alexander Pacek, Benjamin Radcliff, "Turnout and the Vote for Left-of-Centre Parties: A Cross-National Analysis," *British Journal of Political Science,* January 1995, p. 138.

41. Markus M. L. Crepaz, *op. cit.*

42. Arend Lijphart, "Unequal Participation: Democracy's Unresolved Dilemma," *American Political Science Review,* March 1997, vol. 91, no. 1, p. 11.

CHAPTER 13

The New European Parliament

Christopher Lord

Introduction

This review of the formation of the 1999–2004 European Parliament has two purposes. The first is to ask whether the directly elected Parliament should be considered an old or a young institution now that it has reached the age of twenty. Does it represent a mature stage in the parliamentarianization of European Union politics? Or is the EP still an adolescent organization, full of diverse potentials, experimental in its roles, awkward in its practices, incompletely formed, and of uncertain relationship both with its public and other Union institutions? The second and related question is whether the formation of the 1999–2004 Parliament advances our understanding of the exact model of Parliamentary politics that is emerging at Union level. To answer these questions, the chapter will be broken into four sections corresponding to different stages in the formation of a new European Parliament: the electoral link, party-group formation, committee assignments, and the work of the new Parliament in confirming the new Commission in office.

The Electoral Link

Conventional academic analysis might imply that the electoral link is so weak that it can be safely omitted from accounts of the formation of new European Parliaments. The role of the voter is thought to be attenuated by the "second-order" characteristics of a European electoral process dominated by domestic electoral cycles, national party structures, and voter perceptions of the Parliament as a powerless body relative to national democratic institutions. This deprives the EP of an electoral link *ex ante* because voters are not much concerned with the institution that is in fact being elected (or even with the arena in which those elections are taking place). New Parliaments emerge as the cumulative unintended consequence of a series of domestic political games. The electoral link is also missing *ex post*, since MEPs cannot significantly change their chances of reelection through their own political behavior. Choices of partisan alignment in the European Parliament can, therefore, be decoupled from expressions of voter preference without much risk of punishment in future elections.[1]

In contrast, this chapter argues that European elections are of central importance to the formation of each European Parliament. The next section shows that the EP does have a party system capable of translating electoral shifts into changes of political outcome. The argument of this section, however, is that there is an interaction among each five-year round of European elections, the emerging model of European Parliamentary politics, and the way in which new Parliaments set about the task of shaping their own institutional development, whether through Treaty change, internal reorganization of the EP, dealings with the Commission and Council, or reconsideration of relationships with the public.[2] The new European Parliament offers a powerful example. Following turnouts of 65.9 percent in 1979, 63.8 percent in 1984, 62.8 percent in 1989, and 58.5 percent in 1994, voter participation fell to just 49.4 percent in 1999. A slow decline over the period 1979–94, therefore, turned into a sharp fall between 1994–99. Contrary to the prediction that voters would participate in proportion to the power of the European Parliament to affect their lives, turnout since 1979 has been negatively correlated to the expanding role of the EP. For the third time in a row, the 1999 elections offered the voters the opportunity to vote for a Parliament that had recently received additional powers within a political system that was itself expanding its competence. Even more perplexing was why turnout should have fallen so sharply after the European Parliament had demonstrated its capacity to function as a "real parliament" by forcing the resignation of the Commission in March 1999.

To consider how the 1999–2004 Parliament might react to the disjuncture between growing powers and declining electoral legitimation, it is useful to consider some doubts about the manner in which the second-order character of European elections has been specified. Looking at that part of the theory that predicts low participation (rather than voting patterns dominated by national politics), an extensive survey of voters in the 1994 European elections has been used to show that "estimates of the power of the Parliament" do not have much effect.[3] Those voters who believe that the EP lacks power are no less likely to participate than those who perceive it as important. The explanation offered by the authors is that the extent of Parliamentary power is only one of three factors that influence voter participation. The other two are whether those powers are exercised consensually within the Parliament that is being elected, and whether they are shared with other bodies. Where either of these conditions holds, the winner does not take all, the electoral contest is not a "make or break" affair, parties have less incentive to mobilize, and voters less reason to participate. The manner in which the EP has been empowered has only served to confirm the consensual character of the EU's political system; for example, through the requirement that the EP use its main powers only on an absolute majority of its membership. Likewise the EP has not been given power *over* the Commission and Council so much as power *with* those bodies. The Union's legislative procedures allow the EP the opportunity to lever the Commission and Council away from positions they would otherwise have taken. But the Parliament cannot impose itself. It can only coax the Council to accept positions that lie somewhere in between the status quo and the ideal preferences of member states under qualified majority voting; and even within this choice set, the location of the final decision is a matter for negotiation.[4]

Awareness that the empowerment of the EP is insufficient to ensure electoral participation and citizen engagement has provoked the following suggestions for the 1999–2004 Parliament: internal reform to increase public confidence in the integrity of Parliamentary representation; greater differentiation among party groups to demonstrate that meaningful ideological choices are available in the European arena; perpetual readiness to threaten the removal of the Commission; and use of the forthcoming Intergovernmental Conference (IGC) to press for any institutional changes that might facilitate voter turnout and Europeanize the contest. It would, however, be a mistake for the 1999–2004 Parliament to react to its shrinking electoral base by targeting voter mobilization as if it were the sole ingredient of European Parliament legitimacy. If maximizing voter turnout requires a model of Parliamentary politics based on a concentration of power (for example, with the majority in the EP having the right to form the Union's executive), it could have delegitimating effects in a multistate political system that needs to diffuse powers among internally consensual institutions. Conversely, power can be legitimated through the restraints that are put on its use, and not just by the numbers and enthusiasm of those involved in its conferral.[5] The U.S. Congress provides a prime example of how a separation of powers can have legitimating qualities that compensate for mediocre voter participation. The test for the European Parliament is, therefore, to establish an optimal and publicly justified tradeoff among voter mobilization, the diffusion of power, and consensual methods. It cannot be judged on just one of those things.

Party Politics

Another reason for not ignoring the electoral link in the formation of each European Parliament is that domestically obsessed voter behavior need not be a barrier to effective representation where the dimensions of political choice are broadly the same at the national and European levels.[6] For the most part, the party structure of the EP does, indeed, cluster MEPs into left-right groups that correspond to the ranking of voter preferences in the member states where they are elected. The overall party system of the 1999–2004 Parliament thus tilted to the right in response to voter choice. By one calculation, the incoming Parliament has a right-left preponderance of 295 : 273, compared with a left-right lead of 300 : 284 in the outgoing Parliament.[7] This follows a general tendency for the majority in the EP to be of a different political family from that in the Council of Ministers (since European elections often register midterm swings against national governments). Although this is an accidental feature of the Union's institutional design, it does add to the checks and balances within the system. For example, a center right majority in the 1999–2004 Parliament can be expected to temper the domination of the Council by the center-left, which was in government in 13 out of 15 member states at the time of the 1999 elections.

An understanding of the party politics of the new European Parliament requires comparison with its predecessors, which followed a pattern with the following characteristics between 1979 and 1999.[8]

In spite of the national character of European elections, most of the political families represented in European politics have had little difficulty in cohering into transnational Parliamentary groups for the purposes of promoting ideological positions within the EP. Coherence has been high as measured by the frequency with which MEPs from the same group vote together in plenaries.[9] Interview evidence also testifies to a high level of coordination in committees.

Transnational party organization has, however, been easier at some points on the political spectrum than at others. Historical divisions on the center-right between Christian Democrats and Conservatives, made more salient by the issue of European integration itself, have meant that the mainstream right has often been more fragmented than the center left. The far right, on the other hand, has only rarely been able to constitute itself as a group at all, in part because of lack of inclination and internal divisions, and in part because its pariah status reduces the incentive to cohere by making it uncoalitionable with other political families. Likewise, a Parliament that is organized ideologically rather than territorially requires regionalists to choose among distributing themselves among other Europarties and organizing themselves into a group that can at best be only technical in nature.

Coalition building among the parties in the EP has tended to be inclusive. Even some of the smaller party groups have participated in around half of the winning majorities,[10] the only notable exception in the 1994–99 Parliament being the Euroskeptical EDN. One party alignment has, however, tended to stand out in popular perception of the European Parliament, that between the PES and the EPP. The European Parliaments of 1979–99 were characterized by the steady perfection of this relationship through the conclusion of elaborate codes of conduct between the two groups.

In all Parliaments between 1979 and 1999, the PES was the largest group and, therefore, the "senior partner" in the PES–EPP Grand Coalition. Since the end of the 1980s, however, the EPP has sought to improve its position by overcoming the historic divide between the Christian Democratic and Conservative political families on the European center right. The Spanish Partido Popular and the British and Danish Conservatives were brought into the EPP during the 1989–94 Parliament.[11] During the 1994–99 Parliament the EPP further closed the gap of 45 seats at the beginning of the EP to just 13 at the end of the term. The PES–EPP relationship has, therefore, been competitive as well as collusive: Even while the two groups colluded to deploy the powers of the EP, their behavior indicated that relative power (as determined by numbers, coherence, and mobilizing potential) remained important to the locus of compromise and thus the determination of political outcomes.

In previous legislatures the party system of the EP became more concentrated during the course of each Parliament.[12] The tendency for European Parliaments to finish with fewer and larger parties than at the beginning of their five-year terms follows from two patterns that have already been observed. On the one hand, the election of a new Parliament under second-order voting tends to loosen the Europarty system by favoring small parties, protest parties, and new entrants into the political system. On the other, the need to mobilize absolute majorities to exercise the powers of the EP, the PES-EPP duopoly, procedural rules on access to funding, and Parliamentary time and appointments all combine to create structural incentives for national party delegations to migrate from small to large groups.

The party groups that were formed immediately after the elections are illustrated in table 13.1. The 1999–2004 Parliament is the first since 1979 in which the EPP and not the PES has emerged as the largest single party group. In part, this reflects voter choice. But it is also the result of elite-level Parliamentary politics: the decision of the 12 RPR deputies (Sarkozy list) to join the UDF (Bayrou list) in the EPP almost completes the consolidation of the European center right. Apart from the Irish Fianna Fail (blocked by the prior membership of its domestic rival, Fine Gael), the EPP now includes all national parties of the right that are not exclusively Euroskeptical or of the extreme right.

The EPP however, has been able to emerge as the largest Parliamentary group only by practicing an element of flexible integration within its own ranks. In addition to having to accept a name change—to the European People's Party (Christian Democrat) and European Democratic Group—to reflect the now equal standing of Christian Democratic and Conservative Parties in its group, it also had to conclude a special agreement with its second

Table 13.1 The party systems of the 1994–1999 and 1999–2004 European Parliaments
A=Ranking of party groups by size, B=Name of party group, C=Number of seats,
D=Percentage share of seats held by each group, E=Cumulative percentage share of seats
held by groups ranked by size

Start of 1994–99 Parliament (July 1994)					End of 1994–99 Parliament (May 1999)					Start of 1999–2004 Parliament (July 1999)*				
A	B	C	D	E	A	B	C	D	E	A	B	C	D	E
1	PES	198	34.9	34.9	1	PES	214	34.2	34.2	1	EPP	234	37.4	37.4
2	EPP	157	27.7	62.6	2	EPP	201	32.1	66.3	2	PES	180	28.8	66.2
3	ELDR	43	7.6	70.2	3	ELDR	42	6.7	73.0	3	ELDR	50	7.9	74.1
4	GUE	28	4.9	75.1	4	UPE	34	5.4	78.4	4	V	47	7.5	81.6
5	FE	27	4.7	79.8	5	GUE	34	5.4	83.8	5	GUE	42	6.7	88.3
6	RDE	26	4.6	84.4	6	V	27	4.3	88.1	6	TDI	29	4.6	92.9
7	V	23	4.0	88.4	7	ARE	21	3.4	91.5	7	UEN	21	3.4	96.3
8	EN	19	3.3	91.7	8	EN	15	2.4	93.9	8	EDD	16	2.6	98.9
9	ARE	13	2.3	94.0										
	NI	27				NI	38				NI	7		

*The July 1999 figures are as of July 7, 1999.
ARE=European Radical Alliance; ELDR=European Liberal Democrats; EN=Europe of the Nations;
FE=Forza Europa; GUE=Group of the European United Left; NI=Non-attached members;
EPP=European People's Party; PES=Party of European Socialists; RDE=European Democratic Alliance;
TDI=Technical Group of Independent Members; UEN=Union for a Europe of Nations Group;
UDE=Europe of Democracies and Diversities; UPE=Union for Europe; V=Greens.

largest national delegation, the British Conservatives (36 members). This effectively allows the British Conservatives to negotiate opt-outs from EPP group
positions on a case-by-case basis.[13] Given that the British Conservatives had
campaigned on a strongly Euroskeptical platform, it is possible that the EPP
will turn out to be an unstable coalition in the 1999–2004 Parliament. It is
more likely, however, that the terms on which the British Conservatives were
readmitted to the EPP simply indicate that left–right decisions are more important than supranational-intergovernmental ones in the work of the Parliament.
This means that unless a party exists for the sole purpose of protesting against
the existing level of integration, it will tend to align with others in the EP on a
basis of left-right preference. The decision by the British Conservatives to
return to the EPP was accordingly opposed by only 4 of their 36 members.

In contrast to the EPP, the PES enters the 1999–2004 Parliament with
much the same national membership as at the end of the last Parliament.
There will, however, be more equality of influence among its national party
delegations. The commanding position of just one national party (the British
Labour Party with 62 members) will be replaced by a rough balance of influence among delegations from the five largest member states. Beyond affairs
of the EPP and PES, the biggest change is probably the rise of the Green

group, which moved from the seventh largest group in 1994 to the fourth largest in 1999. A further feature is that the new Parliament begins its term with levels of party concentration that were only reached toward the last. 66 percent of MEPs are a part of the two largest groups. For the first time, there are three other groups of between 40 and 50 members (ELDR, GUE, and the Greens), with the result that 88 percent of MEPs are absorbed into the five biggest party formations. The importance of belonging to a transnational party group was, perhaps, most dramatically illustrated by the decision of the Bonino-Panella list from Italy to form a technical group of Independent deputies with the French National Front and Belgian Vlaams Blok, rather than sit as unattached members. Although the 1999–2004 Parliament could yet follow its predecessors by moving toward a still more concentrated party structure during the course of its term, it seems more likely that MEPs have broken with previous practices by rationally anticipating the centripetal tendencies of the Europarty system from the outset.

The most intriguing feature of the new Parliament, however, is that it began with a deliberate attempt to break with "grand coalition politics." One of the clearest expressions of the PES–EPP duopoly had been the practice of dividing the presidency of the Parliament into two terms to be held by leading members of both groups. Following an agreement struck outside the Parliament, the PES felt constrained to insist that its candidate, Mario Soares, preside over the first half of the 1999–2004 parliament. The EPP objected that this failed to acknowledge the verdict of the European elections and its own emergence as the largest group. In preference to a deal with the PES, it accordingly agreed with the ELDR "that the President ... for the first half of the legislature should be Nicole Fontaine [EPP] and for the second half Pat Cox" [ELDR].[14] For its part, the PES failed to negotiate a matching deal with the Greens and the far left.[15] The final result, which gave Fontaine 306 votes, indicated that the EPP and ELDR (with combined membership of 284) picked up some further votes on the right, even though the Pasqua-de Villiers list (12 MEPs) was thought to have abstained.[16]

The leaderships of both the EPP and the ELDR presented their agreement on the presidency as part of a larger challenge to the EPP–PES duopoly, which could well be repeated throughout the course of the Parliament.[17] As an immediate down payment, the agreement also included the assignment of committee chairs and a commitment to a new statute to regulate public standards in the Parliament. Although the agreement was publicly justified as introducing more contestation, political equality, and openness to the Parliament in response to declining voter participation, it was also the product of strategic calculation: the arithmetic of the 1999–2004 Parliament allows the EPP to experiment with long-held ambitions to "develop an effective counter-balance to the Socialists in Europe, and construct a majority political group [of the center–right] in the European Parliament."[18] Coordination with the ELDR would leave the EPP just 30 votes short of an overall majority (of 314). Those further to the right would presumably vote for EPP–ELDR positions in preference to PES ones. Intriguingly, some EPP strategists envisage circumstances in which the Greens might prefer to align with the EPP–ELDR than with the PES, a possibility that the Greens may have been keen to signal by requesting that they be reseated in the center of the Parliament, rather than in their old position to the left of the PES.

Yet, the prospect of regular alignment between the EPP and the ELDR is fraught with difficulty. One problem is the cohesion of the two groups themselves. Choices on a left–right dimension are likely to strain the ELDR, which is divided between left- and right-leaning Liberal Parties. Choices of a supranational-intergovernmental kind are likely to test the EPP, and especially its relationship with the British Conservatives. A further obstacle is the rule that requires an absolute majority of MEPs before most significant powers of the Parliament can be used. Assuming historically average levels of attendance during plenary votes of 65 percent, Simon Hix has calculated that the chances of any group other than the EPP or PES being pivotal to the formation of a winning coalition in the Parliament are minimal:[19] In other words, the EPP–PES combine is part of an iron law of European Parliament politics under present Treaty rules. Legislative coalitions of the EPP and ELDR that are opposed by the PES would almost certainly require the two groups to mobilize 100 percent attendance at plenaries. A final difficulty is that to be effective, the EP has to play a game whose payoffs are set by its relationship with the Commission and the Council. A Council that is dominated by PES governments might prefer high levels of interinstitutional deadlock to accepting Parliamentary amendments that are based on the center right of the EP.

Perhaps the most plausible interpretation of the EPP–ELDR understanding is that it is more likely to restructure the EPP–PES relationship than abolish it. Indeed, far from drawing clearer lines between a Parliamentary majority and opposition, it may presage still grander coalitions, certainly within a PES–EPP–ELDR triangle, and possibly including others too. The leader of the EPP group in the new Parliament, Hans-Goetz Poettering, has, accordingly, stated his intention to chair the Conference of Presidents, which brings together the party leaders to manage the Parliamentary agenda, in a manner that promotes an inclusive pattern of coalition building. Although still very much anchored in consensus politics, such changes could presage greater openness of outcomes: first, if coordination between the EPP and PES survives in a form that ceases to be prior either in time or in political importance to collaboration with other groups; and, second, if the normal politics of EPP–PES collaboration are spiced by the knowledge that either partner could periodically make a unilateral dash for its own preferred solution by mobilizing a full turnout in a plenary, forging temporary alliances with smaller groups, and relying on a measure of dissent in the other large party formation. Although it is the EPP that has taken the lead in opening this possibility, the arithmetic of the 1999–2004 Parliament is especially interesting for the way in which it would support a center-left alternative to EPP–PES collusion as well as a center–right one: A combination of the PES, the ELDR, the GUE, and the Greens would, for example, have 319 votes.

Committee Formation

The process of party-group formation at the beginning of a Parliament is given added urgency by the need for political families and national party delegations to maximize their influence over the membership and chairs of the EP's committees. The EP is a committee-based parliament. Its vast workload can be discharged only by delegation to specialist committees that enjoy political leadership in their areas of competence. Devolution to committees also enables the EP to operate as a "multi-institution," whose powers and functions vary enormously across issues in response to the uneven parliamentarianization of Union politics:[20] Several committees wield the power of legislative codecision with the Council; others have little more than a persuasive role, notably those that cover the second and third pillars (the common foreign and security policy and Justice and Home Affairs); the Budget Committee has some power of the purse; and, most distinctively of all, the Economic and Monetary Committee has the role of examining the European Central Bank and the justifications it offers for its decisions.

What this means is that committee assignments are not only a test of the strength and political skills of the party groups and national party delegations. They are also indications of where actors locate their policy priorities within the complex structure of European Parliamentary representation. The d'Hondt system of proportional representation is used to allocate political opportunities within and among the party groups. The largest party group makes the first nomination to a committee position. Its remaining votes are then reduced in proportion to the number of appointments to be made. Only if it is still the largest group can it make the second nomination, which otherwise passes to the next largest group in the Parliament and so on. Exactly the same process is then used among the national parties inside each group. The assignment of committee chairs for the 1999–2004 Parliament shown in table 13.2 is thus a useful indicator of political strength and policy priorities, both across the EP's parties and across their component national parties.

Table 13.2 Committee chairs for the first half of the new European Parliament, by party group and national party delegation

Committee	Chair	Group	National party
External Affairs	Brok	EPP	Christian Democratische, Germany
Budget	Wynn	PES	Labour, U.K.
Justice/Home Affairs	Watson	ELDR	Liberal, U.K.
Economic and Monetary	Randzio-Plath	PES	Sozial Partei Democratische, Germany
Internal Market	Palacio	EPP	Partido Popular, Spain
Industry	Westendorp	PES	Partido Socialista, Spain
Employment	Rocard	PES	Parti Socialiste, France
Environment	Jackson	EPP	Conservative, U.K.
Agriculture	Graefe zu Baringdorf	Verts	Grüne, Germany
Fishing	Suanzes-Carpenga	EPP	Partido Popular, Spain
Regional Policy	Hatzidakis	EPP	New Democracy, Greece
Culture	Gargani	EPP	Forza Italia, Italy
Development Aid	Miranda	GUE	Partido Comunista, Portugal
Constitutional Affairs	Napolitano	PES	Partito Democratico Sinistra
Womens Rights	Theorin	PES	Labor Party, Sweden
Petitions	Gemelli	EPP	Christiano Democratici, Italy

The Confirmation of the New Commission in Office

Although there are important continuities in the development of the European Parliament since 1979, each of the last three European Parliaments has been a different institution to its predecessor, since regular Treaty changes have redefined the powers of the EP and the manner in which it interacts with the rest of the EU's political system. Until 1993, the third directly elected European Parliament operated under the Single European Act, which gave it powers of legislative amendment under the cooperation procedure (largely in relation to the single market). Until May 1999, the fourth Parliament functioned under the Treaty on European Union (Maastricht). This extended the cooperation procedure to most first-pillar questions, introduced full legislative codecision between the EP and the Council in some limited areas, and gave the EP the right to confirm the new Commission in office. The fifth Parliament will operate under the Amsterdam Treaty. This moves the Union's political system in a more obviously bicameral direction by extending legislative codecision with the Council to about half of first-pillar legislation. It also strengthens the role of the Parliament in confirming the European Commission in office.

Of these two changes, the simplification and extension of codecision is likely to be more important to the 1999–2004 Parliament than the new procedures for confirming the Commission in office. There are structural reasons why the EP is likely to continue to be a law-making parliament, rather than one that trades its legislative independence for a powerful role in the appointment and dismissal of executives. As the EU is a multistate polity whose legitimacy is still widely contested, each of its institutions has to be constructed in a manner that gives continuous and equal influence to representatives of all territorial segments and mainstream political families. This precludes the classic European model in which parliaments have the power to form and dismiss governments in exchange for the prerequisites of stable majoritarianism: external management by the executive authority and internal differentiation into government ("the included") and opposition ("the excluded"). On the other hand, the EU is able to follow the practice of the U.S. Congress, whereby the representative body may function independently of the executive, and legislation is the product of fluid majorities formed on a case-by-case basis. Nonetheless, it is the lesser of the two powers—the confirmation of the new Commission in office—that coincides with the formation of a new European Parliament. Whereas it will be some time before there is adequate data on how the 1999–2004 Parliament is handling the new codecision procedure, its role in the confirmation of the new Commission is already complete and open to analysis.

The investiture of the Santer Commission in 1994–5 confirmed the limits of EP influence over the formation of the EU's governing authority. This was in spite of the following factors that seemed to promise that the Parliament would play an assertive role:

- the alignment of five-yearly Parliamentary terms with those of the Commission
- the Parliament's own success in devising an impressive investiture procedure in which a vote was taken on the Commission presidency, individual Commissioners were examined by Parliamentary committees corresponding to their areas of responsibility, and a vote was then taken on the overall program and composition of the Commission
- the narrowness of the majority (just 18 votes) for the nomination of Jacques Santer as president of the Commission
- the preparedness of the Parliament to write reports that were deeply critical of five of the individual nominees to the College of Commissioners

Yet, the EP was trebly constrained in 1994–5. First, it only had a "take it or leave it" choice to accept or reject the Commission as a whole. Second, the Parliament was perceived as likely to be the first to climb down in any conflict with the European Council over the appointment of a new Commission, given its own dependence on a strong supranational executive, particularly in the legislative process. Third, the investiture of the Santer Commission only confirmed that there was a tradeoff between the power to shape the executive and the independence of a Parliament: As soon as it looked as though the EP might reject the European Council's nominee for the presidency, MEPs from national parties of government came under pressure to vote for Santer, even if that meant breaking with their party group.[21]

However, the 1999–2004 Parliament faced a different structure of political opportunities when it came to the investiture of the Prodi Commission. In part, this was the result of Treaty change. By simultaneously formalizing the Parliament's right to take a separate vote on the Commission presidency while allowing the latter a veto over individual national nominations to the College, Amsterdam gave the EP added bargaining leverage: As well as threatening to reject the overall Commission, it could attempt to influence the role of the president-elect in the selection of individual Commissioners.

A further critical change was that the Prodi Commission was being formed in the shadow of the enforced resignation of its predecessor. This overturned conventional wisdom that the EP would never resort to the "nuclear option" of removing an entire College of Commissioners, not least because the EP had demonstrated that it could bring about such an outcome without going through the formal requirement of organizing a double majority against the Commission (a two-thirds majority of those voting and an absolute majority of the Parliamentary membership)—this could be sufficient to make the Commission's position publicly untenable. The vulnerability of the Commission to Parliamentary attack had, moreover, been steadily increasing well before the removal of the Santer Commission. In 1997 the EP invented a new weapon of public control: By simultaneously passing a suspensory or provisional motion of censure against the Commission while setting up a special committee of inquiry into the BSE crisis, the Parliament succeeded in extracting concessions on the management of food safety, including a transfer of responsibilities within the Commission. As an unelected body with slender resources and little basis for its legitimacy apart from public belief in its competence, the Commission is thought by some insiders to be anxious to avoid even the framing of censure motions. This forces it to cultivate its relations with all sections of the EP, since a censure motion can be tabled by as few as 60 MEPs.[22]

Factors such as these sharpen the incentive for an incoming Commission to use the investiture procedure to win the confidence of the Parliament. The most obvious difficulty in 1999 concerned the political balance of the Prodi Commission. Publication of the final list of national nominations in July 1999 gave the center–left a slight preponderance in the new Commission. The EPP leadership objected that the Commission should reflect the partisan composition of the Parliament and therefore the outcome of European elections. The German Christian Democrats were especially annoyed that both German Commissioners had been assigned to supporters of the ruling coalition, even though the SPD and the Greens had been routed in the 1999 elections. Meanwhile the British Conservatives had been elected on a manifesto that pledged rejection of a new Commission that reappointed any member of the Santer college. To these criticisms of the general shape of the Commission, the hearings of individual Commissioners added some specific objections, notably in the case of the Belgian nominee, Philippe Busquin.

Yet, as with the Santer Commission in 1995, the Prodi Commission was invested without a single change to its composition. From as early as July, it was clear that the German Christian Democrats and the British Conservatives would be unable to unite the EPP around any decision to reject the Prodi Commission on grounds of political balance. The PES was unlikely to reject a College in which it held a majority; and the ELDR made it clear that its priority was the earliest resolution of the institutional crisis on terms that improved standards of public administration. Prodi maintained control over the process of executive formation in two ways. By indicating that the composition of the new Commission had to be sensitive to the political balance of the Council of Ministers as much as to the Parliament, he underlined his own pivotal position in relation to both bodies. By threatening to resign as president-elect when the EP attempted to gain added leverage by subjecting the incoming Commission to two votes of confirmation—one to cover the unexpired period of the 1995–2000 Commission, and the other to invest the 2000–5 Commission in office—he emphasized the Parliament's own need to deploy the confirmation procedure with restraint if it was to secure a swift solution to the institutional crisis.

The investiture of the 1999–2004 Commission was, therefore, another setback for those who saw the confirmation procedure as a means of aligning the EU's executive power with the majority in its Parliament. On the other hand, the procedure paradoxically confirmed the strength of the EP in relation to models of Parliamentary politics other than that of government formation: its importance as a legislator and as a source of nonmajoritarian public control. The extension of codecision to about half of first-pillar legislation gave several Commissioners an interest in using the investiture procedure to form a working relationship with Parliamentary committees in their area of competence. The result was that the process of devising programmatic commitments that had been mainly based on general statements to plenaries in 1994–5 was now more extensively devolved to the Parliament's specialist committees. Each committee issued a long list of detailed questions to which nominee Commissioners had to reply in writing, before being examined in person.[23] The final agreement between Prodi and the EP party leaders (Conference of Presidents) established a continuing obligation on individual Commissioners to attend EP committees in priority to all other engagements, an arrangement that Poettering claimed would amount to a working "Commission in Parliament."

A further feature of the agreement was a promise that the president of the Commission would at least give serious consideration to demanding the resignation of any individual Commissioner who lost a confidence vote in the Parliament. In other words, the investiture of the 1999–2004 may prove to be a landmark in the extension of the EP's public control functions from collective responsibility (dramatically established in March 1999) to individual responsibility.[24] It does, however, need to be noted that the EP failed to extract uniform answers when its committees asked each of the 20 Commissioners to define their area of personal responsibility: Most accepted that they should be held responsible for their own decisions and those of their cabinets, but only some mentioned responsibility for overall policy frameworks and resourcing. Nonetheless, the eventual compromise was an interesting one. Collective and individual responsibility are to be reconciled by the president of the Commission acting as the final arbiter on whether a Parliamentary vote of censure should be grounds to demand the resignation of an individual Commissioner, subject to the Parliament then being able to proceed against the Commission as a whole if it is not satisfied by the president's response. This is likely to confine the EP to nonmajoritarian forms of public control in which Commissioners are only removed for failure to meet agreed standards of public life and not on account of their policy preferences: first, because the president would have strong grounds for rejecting a censure that could be justified only on partisan grounds; and, second, because the Parliament would then need a two-thirds majority against the whole Commission to overturn the president's decision.

Conclusion

To return to the questions that were posed at the beginning of this chapter, the fifth European Parliament clearly represents the maturation of a 20-year process in which practices of transnational representative politics have been steadily perfected: This is true of its powers, party structures, and committee organization. In many ways, the Parliament has also succeeded in replacing the old Commission-Council tandem with a triangular inter-institutional relationship in which the EP often has to be considered as an equal partner of the other two bodies. Even if it lacks the power of government formation, it is a formidable and independent legislator. It is also a source of nonmajoritarian public control in those instances in which it can appeal to agreed standards of public life that go beyond the partisan preferences of particular Parliamentary groups. The least developed aspect of the Parliament remains, however, its relationship with the public. Its electoral-link capacity for representation relies too heavily on a happy coincidence between the left–right character of the

EP's agenda and the principal dimensions of national politics where MEPs are elected, rather than on any direct contribution by the Parliament to the development of a public forum on Union affairs. One suggestion is that there should be more contestation in the politics of the Parliament. However, the logic of consensus is firmly underpinned by inter-institutional relationships within the Union, as well as Treaty rules that constrain the Parliament to employ grand coalition politics if it is to use its powers. The challenge for the 1999–2004 Parliament may, therefore, be to develop a series of "mixed-motive games" in which competition and consensus apply to different stages of the Parliamentary process. The mixture between partnership and threat in its emerging relationship with the Commission would seem to provide one example of such an approach. Another might be to reserve consensus politics for final decisions while introducing more public controversy into the manner in which the Parliament's agenda is set.

Notes

1. The theory was first formulated in K. Reif and H. Schmitt, "Nine Second Order Elections: A Conceptual Framework for the Analysis of European Election Results," *European Journal of Political Research*, vol. 8, no. 3, pp. 3–45, 1980. See also M. Franklin and C. van der Eijk (eds.), *Choosing Europe? The European Electorate and National Politics in the Face of Union*, Ann Arbor: University of Michigan Press, 1996.

2. The work of the European Parliament as a political system builder is intriguingly recounted in R. corbett, *The European Parliament's Role in Closer European Integration*, Basingstoke: Macmillan, 1998.

3. J. Blondel, R. Sinnott and P. Svensson, *People and Parliament in the European Union: Participation, Democracy and Legitimacy*, Oxford: Clarendon Press, 1998, p. 255.

4. For analysis of the conditional character of European Parliamentary power, see especially G. Tsebelis, "The Power of the European Parliament as a Conditional Agenda-Setter," *American Political Science Review*, vol. 88, no. 1, pp. 128–42, 1994. See also G. Tsebelis and A. Kreeppl, "The History of Conditional Agenda-Setting in European Institutions," *European Journal of Political Research*, vol. 33, no. 1, pp. 43–71, 1998.

5. The argument that Union institutions should concentrate on substitutes to legitimation by mass representative process is made in A. Héritier, "Elements of Democratic Legitimation in Europe: An Alternative Perspective," in *Journal of European Public Policy*, vol. 6, no. 2, pp. 269–82, 1999.

6. See M. Marsh and P. Norris (eds.) *Political Representation in the European Parliament*, special issue of the *European Journal of Political Research*, vol. 32, no. 2, 1997.

7. These figures assume that the ELDR and the Bonino/Panella list from Italy are capable of aligning with either the left or the right depending on the issue, and that the Danish Euroskeptical parties belong to the left.

8. A full analysis of the party system in the European Parliament is to be found in S. Hix and C. Lord, *Political Parties in the EU*, London: Macmillan, 1997.

9. Roll-call analysis of voting in the European Parliament was pioneered in F. Attinà, "The Voting Analysis of the European Parliament Members and the Problem of Europarties," *European Journal of Political Research*, vol. 18, no. 3, pp. 557–579, 1990. Further data is to be found in various sources: Hix and Lord, *op. cit.*, T. Raunio, *Party Group Behaviour in the European Parliament*, Tampere: University of Tampere, 1996.

10. S. Hix and C. Lord, *op. cit.*, pp. 134–5.

11. For a full account of the rapprochement between the Christian Democratic and Conservative Parties, see T. Jansen, *The European People's Party: Origins and Development*, New York: St. Martin's Press, 1998. Also T. Jansen, "The Integration of the Conservatives into the European People's Party," in D. Bell and C. Lord (eds.), *Transnational Parties in the European Union*, Aldershot: Ashgate, 1998.

12. This pattern was first identified in L. Bardi, "Transnational Trends in European Parties and the 1994 Elections of the European Parliament," *Party Politics*, vol. 2, no. 1, pp. 585–610.

13. Author's conversation with the leader of the British Conservatives in the European Parliament, Edward Macmillan-Scott, MEP.

14. ELDR press release, "Constitutive Agreement—Not a Political Coalition," Brussels, July 15, 1999.

15. "Nicole Fontaine fait le plein des voix de droite," *Libération*, July 21, 1999.

16. Ibid.

17. See ELDR press release.

18. Quoted from G. Rinsche and K. Welle, *Mehrheit für die Mitte: Muß das Europäische Parlament socialistisch sein?* Bonn: Konrad Adenauer Stiftung, 1999.

19. S. Hix, *The Political System of the European Union*, London: Macmillan, 1999, p. 82. Calculations based on Shapley-Shubik power index.

20. The notion that the EU is differentially parliamentrianized across issues is a variation on Paul Magnette's description of the Union as "un régime semi-parlementaire." P. Magnette, "L'Union Européenne: Un Régime Semi-Parlementaire," in P. Delwit, J.-M. De Waele, and P. Magnette, *À quoi sert le Parlement Européen*, Brussels: Éditions Complexe, 1999.

21. See S. Hix and C. Lord, "The Making of a President: The European Parliament and the Confirmation of Jacques Santer as President of the Commission," *Government and Opposition*, vol. 31, no. 1, pp. 62–76.

22. Based on a conversation with Martin Westlake. This is further explained in C. Lord, *Democracy in the European Union*, Sheffield: UACES/Sheffield University Press, 1999.

23. The Parliament published the written answers of each Commissioner to its committees in full, together with transcripts of the subsequent hearings.

24. For details of what seems to have been agreed between President Prodi and the EP's party leaders, see speeches by Poettering and Cox in the European Parliament debate of September 14, 1999. *Compte rendu in extenso des séances*, 14-09-1999, Luxembourg, Parlement Européen.

Appendices

Laurent de Boissieu, Françoise Chauve, Jean Chiche, Pascal Perrineau, Christian Vandermotten

Tables A1 (A1.1 to A1.15) Results of the European elections in the 15 countries of the European Union (June 1999)

Table A1.1 Austria

	1999		
Registered voters	5,847,605		
Voters	2,888,733		
Votes counted	2,801,353		
	Votes	*% V.C.*	*Seats*
Sozialdemokratische Partei Österreichs (SPÖ)	888,338	31.71	7
Österreichische Volkspartei (ÖVP)	859,175	30.67	7
Freiheitliche Partei Österreichs (FPÖ)	655,519	23.4	5
Die Grünen/Die Grüne Alternative (GRÜNE)	260,273	9.29	2
Liberales Forum (LIF)	74,467	2.66	–
Christlich Soziale Allianz (CSA)	43,084	1.54	–
Kommunistische Partei Österreichs (KPÖ)	20,497	0.73	–

Table A1.2 Belgium

<table>
<tr><td></td><td colspan="4" align="center">1999</td></tr>
<tr><td align="right">Registered voters</td><td>7,343,466</td><td></td><td></td><td></td></tr>
<tr><td align="right">Voters</td><td>6,686,222</td><td></td><td></td><td></td></tr>
<tr><td align="right">Votes counted</td><td>6,223,166</td><td></td><td></td><td></td></tr>
<tr><td></td><td>Votes</td><td>% Votes counted/college</td><td>Total % votes counted</td><td>Seats</td></tr>
<tr><td>Flemish-speaking electoral college (14 seats)</td><td>3,872,431</td><td>100,00</td><td>62,23</td><td></td></tr>
<tr><td>Vlaamse Liberalen en Democraten (VLD)</td><td>847,103</td><td>21,88</td><td>13,61</td><td>3</td></tr>
<tr><td>Christelijke Volkspartij (CVP)</td><td>839,688</td><td>21,68</td><td>13,49</td><td>3</td></tr>
<tr><td>Vlaams Blok (VB)</td><td>584,392</td><td>15,09</td><td>9,39</td><td>2</td></tr>
<tr><td>Socialistische Partij (SP)</td><td>550,237</td><td>14,21</td><td>8,84</td><td>2</td></tr>
<tr><td>Volksunie (VU)/Ideeën voor de 21ste Eeuw (ID 21)</td><td>471,239</td><td>12,17</td><td>7,57</td><td>2</td></tr>
<tr><td>AGALEV-De Vlaamse Groenen</td><td>464,043</td><td>11,98</td><td>7,46</td><td>2</td></tr>
<tr><td>VIVANT</td><td>67,107</td><td>1,73</td><td>1,08</td><td>–</td></tr>
<tr><td>Partij van de Arbeid-Antifascistische Enheid (PvdA-AE)</td><td>21,998</td><td>0,57</td><td>0,35</td><td>–</td></tr>
<tr><td>Partij voor Nieuwe Politiek België (PNPB)</td><td>17,095</td><td>0,44</td><td>0,27</td><td>–</td></tr>
<tr><td>Sociaal-Liberalen Democraten (SOLIDE)</td><td>9,529</td><td>0,25</td><td>0,15</td><td>–</td></tr>
<tr><td>French-speaking electoral college (10 seats)</td><td>2,313,835</td><td>100</td><td>37,18</td><td></td></tr>
<tr><td>Parti Réformateur Libéral (PRL)</td><td></td><td></td><td></td><td></td></tr>
<tr><td>Front Démocratique des Francophones (FDF)</td><td></td><td></td><td></td><td>2</td></tr>
<tr><td>Mouvement des Citoyens pour le Changement (MCC)</td><td>624,444</td><td>26.99</td><td>10.03</td><td>1</td></tr>
<tr><td>Parti Socialiste (PS)</td><td>596,565</td><td>25.78</td><td>9.59</td><td>3</td></tr>
<tr><td>ECOLO</td><td>525,313</td><td>22.70</td><td>8.44</td><td>3</td></tr>
<tr><td>Parti Social Chrétien (PSC)</td><td>307,909</td><td>13.31</td><td>4.95</td><td>1</td></tr>
<tr><td>Front National (FN)</td><td>94,848</td><td>4.10</td><td>1.52</td><td>–</td></tr>
<tr><td>VIVANT</td><td>55,135</td><td>2.38</td><td>0.89</td><td>–</td></tr>
<tr><td>DEBOUT</td><td>46,089</td><td>1.99</td><td>0.74</td><td>–</td></tr>
<tr><td>Parti Communiste de Belgium (PCB)</td><td>25,541</td><td>1.10</td><td>0.41</td><td>–</td></tr>
<tr><td>Front Nouveau de Belgium (FNB)</td><td>24,792</td><td>1.07</td><td>0.40</td><td>–</td></tr>
<tr><td>Parti pour une Nouvelle Politique Belge (PNPB)</td><td>10,479</td><td>0.45</td><td>0.17</td><td>–</td></tr>
<tr><td>Parti Communautaire National-Européen (PCN-E)</td><td>2,720</td><td>0.12</td><td>0.04</td><td>–</td></tr>
<tr><td>German-speaking electoral college (1 seat)</td><td>36,900</td><td>100</td><td>0.59</td><td>1</td></tr>
<tr><td>Christliche Soziale Partei-Europaïsche Volkspartei (CSP-EVP)</td><td>13,456</td><td>36.47</td><td>0.22</td><td>–</td></tr>
<tr><td>Partei für Freiheit und Fortschritt (PFF)</td><td>7,234</td><td>19.60</td><td>0.12</td><td>–</td></tr>
<tr><td>ECOLO</td><td>6,276</td><td>17.01</td><td>0.10</td><td>–</td></tr>
<tr><td>Sozialistische Partei (SP)</td><td>4,215</td><td>11.42</td><td>0.07</td><td>–</td></tr>
<tr><td>Partei der Deutschsprachigen Belgier (PDB)</td><td>3,661</td><td>9.92</td><td>0.06</td><td>–</td></tr>
<tr><td>VIVANT</td><td>1,198</td><td>3.25</td><td>0.02</td><td>–</td></tr>
<tr><td>JUROPA</td><td>788</td><td>2.14</td><td>0.01</td><td>–</td></tr>
<tr><td>Partei der Arbeit Belgiens-Antifaschistische Einheit (PAB-AE)</td><td>72</td><td>0.20</td><td>0.00</td><td>–</td></tr>
</table>

Table A1.3 Denmark

	1999		
Registered voters	4,012,440		
Voters	2,021,922		
Votes counted	1,969,121		

	Votes	% V.C.	Seats
Venstre, Danmarks Liberale Parti (V)	459,524	23.34	5
Socialdemokratiet (SD)	324,193	16.46	3
JuniBevaegelsen	317,238	16.11	3
Det Radikale Venstre (RV)	180,073	9.14	1
Det Konservative Folkeparti (KF)	166,793	8.47	1
Folkebevaegelsen mod EU	143,678	7.30	1
Socialistik Folkeparti (SF)	140,178	7.12	1
Dansk Folkeparti (DF)	114,449	5.81	1
Centrum-Demokraterne (CD)	68,530	3.48	–
Kristeligt Folkeparti (KRF)	40,240	2.04	–
Fremskridtspartiet (FRP)	14,225	0.72	–

Table A1.4 Finland

	1999		
Registered voters	4,141,098		
Voters	1,248,122		
Votes counted	1,242,303		

	Votes	% V.C.	Seats
Kansallinen Kokoomus (KOK)	313,960	25.27	4
Suomen Keskusta (KESK)	264,640	21.30	4
Suomen Sosialidemokraattinen Puolue (SDP)	221,836	17.86	3
Vihreä Liitto (VIHR)	166,786	13.42	2
Vasemmistopoliito (VAS)	112,757	9.07	1
Svenska Folkpartiet (RKP-SFP)	84,153	6.77	1
Suomen Kristillinen Liitto (SKL)	29,637	2.38	1
Kirjava "Puolue"-Elonkehän Puolesta (KIPU)*	29,215	2.35	
Perussuomalaiset (PS)*	9,854	0.79	
Suomen Kommunistinen Puolue (SKP)	7,556	0.60	
Eläkeläiset Kansan Asialla (EKA)*	1,909	0.15	

'KIPU, PS, and EKA formed an electoral alliance.

Table A1.5 France

	Metropolitan France			Overseas			Total			Seats
	Votes	% Registered voters	% counted	Votes	% Registered voters	% counted	Votes	% Registered voters	% counted	
Registered voters	38,590,748			1,541,769			40,132,517			
Abstentions	20,163,808	52.25		1,202,554	78.00		21,366,362	53.24		
Voters	18,426,940	47.75		339,215	22.00		18,766,155	46.76		
Blanks and voids	1,094,899	2.84		24,084	1.56		1,118,983	2.79		
Votes Counted	17,332,041	44.91		315,131	20.44		17,647,172	43.97		
LO-LCR list (Laguiller)	907,014	2.35	5.23	7,797	0.51	2.47	914,811	2.28	5.18	5
Parti Communiste list (Hue)	1,184,897	3.07	6.84	11,594	0.75	3.68	1,196,491	2.98	6.78	6
Parti Socialiste list (Hollande)	3,791,820	9.83	21.88	82,411	5.35	26.15	3,874,231	9.65	21.95	22
L'Ecologie, Les Verts list (Cohn-Bendit)	1,691,441	4.38	9.76	24,288	1.58	7.71	1,715,729	4.28	9.72	9
UDF list (Bayrou)	1,602,595	4.15	9.25	36,404	2.36	11.55	1,638,999	4.08	9.29	9
RPR-DL union list (Sarkozy)	2,173,215	5.63	12.54	89,986	5.84	28.56	2,263,201	5.64	12.82	12
Rassemblement pour la France list (Pasqua)	2,279,233	5.91	13.15	25,311	1.64	8.03	2,304,544	5.74	13.06	13
Moins d'impôts maintenant list (Miguet)	305,790	0.79	1.76	6,660	0.43	2.11	312,450	0.78	1.77	
Vivant énergie France list (Maudrux)	123,290	0.32	0.71	1,271	0.08	0.40	124,561	0.31	0.71	
Mouvement national list (Mégret)	572,727	1.48	3.30	6,110	0.40	1.94	578,837	1.44	3.28	
Front national list (Le Pen)	995,588	2.58	5.74	9,697	0.62	3.08	1,005,285	2.50	5.70	5

Ligue nationaliste list (Guerrin)	683	0.00	0.00		0	0	683	0.00	0.00	6
Chasse, Pêche, nature, traditions list (Saint-Josse)	1,194,720	3.10	6.89	1,143	0.07	0.36	1,195,863	2.98	6.78	
Mouvement écologiste indépendant list (Waechter)	268,038	0.69	1.55		0	0	268,038	0.67	1.52	
Combat pour l'emploi list (Larrouturou)	173,650	0.45	1.00	4,414	0.29	1.40	178,064	0.44	1.01	
Parti de la loi naturelle list (Frappé)	65,071	0.17	0.38	6,338	0.41	2.01	71,409	0.18	0.40	
MI OU, MI MWEN list (Jos)	0	0	0	1,707	0.11	0.54	1,707	0.00	0.01	
Politique de vie pour l'Europe list (Cotten)	274	0.00	0.00	0	0	0	274	0.00	0.00	
Parti humaniste list (Chanut-Sapin)	1,995	0.01	0.01	0	0	0	1,995	0.00	0.01	
Vive le fédéralisme list (Allenbach)	0	0	0	0	0	0	0	0	0	

Source: Journal officiel June 22, 1999.

Table A1.6 Germany

	1999		
Registered voters	60,786,904		
Voters	27,468,932		
Votes counted	27,058,273		
	Votes	*% V.C.*	*Seats*
Christlich Demokratische Union Deutschlands (CDU)	10,628,224	39.28	43
Sozialdemokratische Partei Deutschlands (SPD)	8,307,085	30.70	33
Christlich Soziale Union in Bayern (CSU)	2,540,007	9.39	10
Bündnis 90/Die Grünen (GRÜNE)	1,741,494	6.44	7
Partei des Demokratischen Sozialismus (PDS)	1,567,745	5.79	6
Freie Demokratische Partei (FDP)	820,371	3.03	–
Die Republikaner (REP)	461,038	1.70	–
Die Tierschutzpartei-Mensch Umwelt Tierschutz (TIERSCHUTZ)	185,186	0.68	–
Die Grauen-Graue Panther (GRAUE)	112,142	0.41	–
Nationaldemokratische Partei Deutschlands (NPD)	107,662	0.40	–
Ökologisch-Demokratische Partei (ÖDP)	100,048	0.37	–
Die Frauen-Feministische Partei (DIE FRAUEN)	100,128	0.37	–
Autofahrer-und Bürgerinteressen Partei Deutschlands (APD)	97,984	0.36	–
Partei der Arbeitslosen und Sozial Schwachen (PASS)	71,430	0.26	–
Partei Bibeltreuer Christen (PBC)	67,732	0.25	–
Naturgesetz Partei, Aufbruch zu neuem Bewusstsein (NATURGESETZ)	38,139	0.14	–
Automobile Steuerzahler Partei (ASP)	34,029	0.13	–
Christliche Mitte für ein Deutschland nach Gottes Geboten (CM)	30,746	0.11	–
Bayernpartei (BP)	14,950	0.06	–
Humanistische Partei (HP)	11,505	0.04	–
Bürgerrechtsbewegung Solidarität (BüSo)	9,431	0.03	–
Deutsche Zentrumspartei (ZENTRUM)	7,080	0.03	–
Familien Partei Deutschlands (FAMILIE)	4,117	0.02	–

Table A1.7 Great Britain

	1999		
Registered voters	44,495,741		
Voters*			
Votes counted	10,681,082		
	Votes	*% V.C.*	*Seats*
Conservative Party (CP)	3,578,217	33.50	36
Labour Party (LAB)	2,803,821	26.25	29
Liberal Democrat Party (LD)	1,266,549	11.86	10
Independence Party (UKIP)	696,057	6.52	3
Green Party (GP)	625,378	5.86	2
Scottish National Party (SNP)	268,528	2.51	2

Table A1.7 (Continued)

	1999		
	Votes	% V.C.	Seats
Democratic Unionist Party (DUP)	192,762	1.80	1
Social Democratic and Labour Party (SDLP)	190,731	1.79	1
Plaid Cymru (PC)	185,235	1.73	2
Pro-European Conservative Party (PECP)	138,097	1.29	–
Ulster Unionist Party (UUP)	119,507	1.12	1
Sinn Fein (SF)	117,643	1.10	–
British National Party (BNP)	102,647	0.96	–
Liberal Party	93,051	0.87	–
Socialist and Labour Party (SLP)	86,749	0.81	–
Independent	57,909	0.54	–
Scottish Socialist Party (SSP)	39,720	0.37	–
Alternative Labour list (supported by left Alliance) (Alab)	26,963	0.25	–
Progressive Unionist Party of Northern Ireland	22,494	0.21	–
Natural Law Party (NLP)	21,327	0.20	–
United Kingdom Unionist Party (UKU)	20,283	0.19	–
Alliance Party of Northern Ireland (All)	14,391	0.13	–
Socialist Alliance (SAll)	7,203	0.07	–
Humanist Party (HP)	2,586	0.02	–
Weekly Worker (WW)	1,724	0.02	–
Socialist Party of Great Britain (SPGB)	1,510	0.01	–

*In the United Kingdom, blank and void votes are not accounted for. We therefore do not have a total number of "Voters" that includes "votes counted" and blank or void votes. The percentage of the latter is estimated to be less than 0.5 percent of registered voters.

Table A1.8 Greece

	1999		
Registered voters	9,387,417		
Voters	6,711,728		
Votes counted	6,427,648		
	Votes	% V.C.	Seats
Nea Dimokratia (ND)	2,313,265	35.99	9
Panellinio Socialistiko Kinima (PASOK)	2,116,507	32.93	9
Kommounistiko Komma Elladas (KKE)	557,134	8.67	3
Dimokratiko Kinoniko Kinima (DIKKI)	440,079	6.85	2
Synaspismos tis aristeras kai tis proodou (SYN)	332,116	5.17	2
Politiki Anixi (POLAN)	146,760	2.28	
I Fileleftheri-Stefanos Manos (FIL)	103,850	1.62	
Komma Ellinon Kynigon	64,063	1.00	
Enosi Kentroon	52,457	0.82	
Proti Grammi	48,726	0.76	
Kollatos Anexartito Politiko Kinima-Ecologiko Elliniko	42,539	0.66	
Dimosthenis Vergis-Ellines Ecologi	30,741	0.48	
Rizospastiki Antiapagoreftiki Kinisi (RAK)	21,369	0.33	
Marxistiko-Leninistiko Kommounistiko Komma Elladas (M-L. KKE)	16,786	0.26	

Table A1.8 (Continued)

	1999		
	Votes	% V.C.	Seats
Komma Ellinismou	16,715	0.26	
Ecologi Enallaktiki-Hatzipanagiotou	14,822	0.23	
Lefko	14,256	0.22	
Kinima Anergon Ellinon	13,886	0.22	
Metopo Rizopastikis Aristeras	10,930	0.17	
Komma Ellinon Fileleftheron	10,377	0.16	
Aristera	9,697	0.15	
Agonistiko Socialistiko Komma Elladas (ASKE)	9,543	0.15	
Socialistiko Ergatiko Komma-Ergatiki Allileggyi	8,049	0.13	
Elliniko Metopo	7,781	0.12	
Ouranio Toxo-Evropaïko Kinima	4,983	0.08	
Organosi gia tin anasvngrotisi tou KKE (OAKKE)	4,622	0.07	
Aftodynamo Kinima Ethniki Politikis (AKEP)	4,260	0.07	
Olympismos	4,152	0.06	
Komma Fysikou Nomou	2,812	0.04	
Christopistia	2,153	0.03	
Anthropinon Dikaiomaton	1,255	0.02	
Ecologi Enallaktiki-Schizas	466	0.01	
Anexartiti Politiki Parousia	105	0.00	
Dimokratiki Frontida	90	0.00	
Ethnikon Agoniston	74	0.00	
Diakommatiki Kinisi	57	0.00	
Athinaïki Dimokratia	49	0.00	
Elliniki Pangosmia Olympiaki Dimokratia (EPODI)	43	0.00	
Kinonoki Merimna	30	0.00	
Ethniko Patriotiko Komma (EPAK)	28	0.00	
Ethniki Patriotiki Topiki Agonistiki Kinisi Allinon (EPTAKKE)	21	0.00	

Table A1.9 Ireland

	1999		
Registered voters	2,864,361		
Voters	1,438,287		
Votes counted	1,391,740		

	Votes	% V.C.	Seats
Fianna Fail (FF)	537,757	38.63	6
Fine Gael (FG)	342,171	24.59	4
Labour Party (Lab)	121,542	8.73	1
Green Party (GP)	93,100	6.69	2
Sinn Fein	88,165	6.33	–
Independent (Pat Cox)	63,954	4.60	1
Independent (Dana)	51,086	3.67	1
Other	93,965	6.75	–

Table A1.10   Italy

	1999		
Registered voters	49,333,017		
Voters	34,910,815		
Votes counted	30,985,217		

	Votes	% V.C.	Seats
Forza Italia (FI)	7,829,442	25.27	22
Democratici di Sinistra (DS)	5,395,287	17.41	15
Alleanza Nazionale (AN) + Patto Segni (PS)	3,202,821	10.34	9
E. Bonino: Riformatori list	2,631,118	8.49	7
Democratici-Prodi/La Rete (I)	2,407,918	7.77	7
Lega Nord (LN)	1,395,535	4.50	4
Partito Rifondazione Comunista (RC)	1,328,491	4.29	4
Partito Popolare Italiano-Popolari (PPI)	1,319,484	4.26	4
Centro Cristiano Democratico (CCD)	806,422	2.60	2
Socialisti Democratici Italiani (SDI)	671,820	2.17	2
Cristiano Democratici Unitari (scission PPI) (CDU)	670,063	2.16	2
Partito dei Comunisti Italiani (scission PRC) (CI)	622,252	2.01	2
Federazione dei Verdi (VERDI)	548,899	1.77	2
Unione Democratici Europei (U.D. EUR)	499,498	1.61	1
Movimento Sociale Fiamma Tricolore (scission AN) (MS)	371,042	1.20	1
Rinnovamento Italiano-DINI (RI)	353,805	1.14	1
Partito Pensionati (PENSIO)	232,166	0.75	1
Partito Republicano Italiano/Liberali (PRI)	168,178	0.54	1
Südtiroler Volkspartei (SVP)	155,749	0.50	–
Liga Veneta Union (LVU)	118,101	0.38	–
Lega d'Azione Meridonale-Lista Cito (CITO)	93,353	0.30	–
Partito Sardo d'Azione/Partito Consumatori (PSdA/PC)	60,534	0.20	–
Partito Socialista (SOC)	42,544	0.14	–
Union Valdôtaine (UV)	41,227	0.13	–
Partito Umanista (PU)	15,089	0.05	–
Cobas per l'Autorganizzazione (COBAS)	4,369	0.01	–

Table A1.11   Luxembourg

	1999
Registered voters	233,602
Voters	202,384
Votes counted	184,898
Grand total of votes counted	1,013,783

	Votes for list	% V.C.	Seats
Chrëschtlech Sozial Volkspartei (CSV)	167,868	31.70	2
Lëtzebuerger Sozialistesch Aarbechterpartei (LSAP)	134,874	23.60	2
Demokratesch Partei (DP)	121,050	20.50	1

Table A1.11 (Continued)

	1999		
	Votes for list	% V.C.	Seats
Déi Gréng (Verts)	70,038	10.70	1
Aktiounskomitee fir Demokratie a Rentegerechtegkeet (ADR)	64,860	9.00	
Déi Lenk-La Gauche	20,448	2.80	
Gréng a Liberal Allianz (GaL)	9,516	1.80	

*Voting in Luxembourg is by national proportional representation by the d'Hondt system. Each voter has as many votes as there are seats to be attributed (six). He can mix his votes. The voter can either attribute all of his votes to a single list, distribute them over different lists, or distribute them among candidates of the same list.

Table A1.12 Netherlands

	1999		
Registered voters	11,862,864		
Voters	3,560,764		
Votes counted	3,544,408		
	Votes	% V.C.	Seats
Christen Democratisch Appèl (CDA)	954,898	26.94	9
Partij van de Arbeid (PvdA)	712,929	20.11	6
Volkspartij voor Vrijheid en Democratie (VVD)	698,050	19.69	6
Groenlinks	419,869	11.84	4
Staatkundig Gereformeerde Partij (SGP)			1
Gereformeerd Politiek Verbond (GPV)			1
Reformatorische Politieke Federatie (RPF)	309,612	8.74	1
Democraten 66 (D 66)	205,623	5.80	2
Socialistische Partij (SP)	178,642	5.04	1
De Europese Partij (DEP)	23,231	0.66	
CentrumDemocraten (CD)	17,740	0.50	
Europese Verkiezers Platform Nederland (EVPN)	13,234	0.37	
Independent	10,580	0.30	

Table A1.13 Portugal

	1999		
Registered voters	8,600,643		
Voters	3,465,301		
Votes counted	3,351,675		
	Votes	% V.C.	Seats
Partido Socialista (PS)	1,491,963	44.51	12
Partido Social Democrata (PPD-PSD)	1,077,665	32.15	9
Coligaçao Democratica Unitaria (CDU)[1]	357,575	10.67	2
Partido Popular (CDS-PP)	282,928	8.44	2

Table A1.13   (Continued)

	1999		
	Votes	*% V.C.*	*Seats*
Bloco de Esquerda (BE)[2]	62,022	1.85	
Partido Comunista dos Trabalhores Portugueses (PCTT-MRPP)	30,358	0.91	
Partido Popular Monarquico (PPM)	16,164	0.48	
Movimento o Partido da Terra (MPT)	13,669	0.41	
Partido de Solidariedade Nacional (PSN)	8,764	0.26	
Partido Operario de Unidade Socialista (POUS)	5,508	0.16	
Partido Democratico do Atlântico (PDA)	5,059	0.15	

[1] Partido Comunista Português (PCP)/Partido ecologista Os Verdes (PEV).
[2] Uniao Democratica Popular (UDP)/Partido Socialista Revolucionario (PSR)/Politica XXI.

Table A1.14   Spain

	1999		
Registered voters	32,944,451		
Voters	21,209,685		
Votes counted	20,684,196		
	Votes	*% V.C.*	*Seats*
Partido Popular (PP)	8,364,767	40.44	27
Partido Socialista Obrero Español/Progresistas (PSOE-Prog)[1]	7,420,035	35.87	24
Izquierda Unida-Esquerra Unida I Alternativa (IU-EUIA)[2]	1,213,254	5.87	4
Convergencia i Unio (CiU)[3]	934,259	4.52	1
			2
Coalicion Europea (CE)[4]	676,287	3.27	2
Coalicion Nacionalista/Europa de los Pueblos (CN-EP)[5]	611,801	2.96	2
Bloque Nacionalista Galego (BNG)	347,205	1.68	1
Euskal Herritarrok (EH)	306,508	1.48	1
Coalicion Electoral "Los Verdes" las Izquierdas de los Pueblos[6]	299,851	1.45	
Los Verdes-Grupo Verde (LV-GV)	137,038	0.66	
Union Centrista/Centro Democratico y Social (UC/CDS)	41,496	0.20	
Union del Pueblo Leonés (UPL)	33,394	0.16	
Confederacion de Organizaciones Feministas (COFEM-FEMEK)	27,497	0.13	
Partido Comunista de los Pueblos de Espana (PCPE)	25,304	0.12	
Union Renovadora Asturiana (URAS)	22,405	0.11	
Partit per la Independencia (INDEP)	17,810	0.09	
Extremadura Unida (EU)	15,713	0.08	
Falange Española Independiente (FEI)	15,579	0.08	
Partido Democrata Espanol (PADE)	15,350	0.07	
Partiu Asturianista (PAS)	15,189	0.07	
Tierra Comunera-Partido Nacionalista Castellano (TC-PNC)	13,119	0.06	
Alianza por la Unidad Nacional (AUN)	12,318	0.06	
Partido Humanista (PH)	11,999	0.06	
La Falange (FE)	10,490	0.05	
Socialistas Independientes de Extremadura (SIEX)	10,086	0.05	

Table A1.14 (Continued)

| | 1999 | | |
	Votes	% V.C.	Seats
Unidad regionalista de Castilla y Leon (URCL)	9,653	0.05	
Partido de la Ley Natural (PLN)	8,591	0.04	
Partido de los Autonomos de España y de las Agrupaciones Independientes Espanolas (PAE-I)	8,272	0.04	
Asamblea de Andalucia (A)	8,220	0.04	
Alternativa Comunidad Valenciana (ACV)	7,943	0.04	
Andecha Astur (AA)	7,933	0.04	
Democracia Nacional (DN)	7,785	0.04	
Coalicion "Union de Regiones" (UDR)[7]	7,526	0.04	
Coalicion Extremena PREX-CREX	7,271	0.04	
Salamanca, Zamora, Leon Prepal (PREPAL)	6,718	0.03	
Coalicion "Por la derogacion del Tratado de Maastricht" (DM)	5,530	0.03	

[1]Partido Socialista Obrero Español (PSOE)/Partit dels Socialistes de Catalunya (PSC)/Partido Democratico de la Nueva Izquierda (PDNI, ex IU).
[2]Partido Comunista de España (PCE)/Partido de Accion Socialista (PASOC)/Izquierda Republicana (IR)/etc.
[3]Convergència Democratica de Catalunya (CDC)/Unio Democratica de Catalunya (UDC).
[4]Coalicion Canaria (CC)/Partido Andalucista (PA)/Unio Valenciana (UV)/Partido Aragonés (PAR).
[5]Eusko Alderdi Jeltzalea-Partido Nacionalista Vasco (EAJ-PNV)/Eusko Alkartasuna (EA)/Esquerra Republicana de Catalunya (ERC)/Unio Mallorquina (UM)/Els Verds-Confederacio ecologista de Catalunya.
[6]Iniciativa per Catalunya (IC)/Los Verds/Chunta Aragonesista/Esquerda de Galicia/Izquierda Andaluza.
[7]Partido Regional Independentista Madrileno (PRIM)/Union regionalista Almeriense (URAL)/Union Centre Balear (UCB)/Partido de Gran Canaria (PGC)/Partido regionalista de Castilla-La Mancha (PRCM)/Union del Pueblo Balear (UPB).

Table A1.15 Sweden

		1999		
Registered voters	6,664,205			
Voters	2,588,514			
Votes counted	2,529,437			

	Votes	% V.C.	Seats
Arbetarpartiet-Socialdemokraterna (SOC)	657,497	25.99	6
Moderata Samlingspartiet (MSP)	524,755	20.75	5
Vänsterpartiet (VP)	400,073	15.82	3
Folkpartiet Liberalerna (FP)	350,339	13.85	3
Miljöpartiet de Gröna (MP)	239,946	9.49	2
Kristdemokraterna (KDS)	193,354	7.64	2
Centerpartiet (CP)	151,422	5.99	1
Sverigedemokraterna (SD)	8,568	0.34	–
Rättvisepartiet Socialisterna	1,430	0.06	–
Partiet för Naturens Lag	435	0.02	–
Europeiska Arbetarpartiet (AEP)	99	0.00	–
Allianspartiet	75	0.00	–

Table A1.15 (Continued)

	1999		
	Votes	% V.C.	Seats
Hej Du Partiet	72	0.00	–
Vikingspartiet Sverige ut ur EU	64	0.00	–
Strandskyddspartiet	47	0.00	–
Republikanerna	33	0.00	–
Kommunistika Förbundet	32	0.00	–
Oberoende Liberaler	24	0.00	–
Älta Skol-Och Idrottsparti	21	0.00	–
Crisis	3	0.00	–
Dust	3	0.00	–
Other Parties	1,125	0.04	–

Tables A2 (A2.1 to A2.15) Evolution in the results of the European elections in the 15 countries of the European Union, 1979–1999

Table A2.1 Austria

Parties	1996			1999		
Registered voters	5,800,377			5,847,660		
Voters	3,928,538		67.7	2,888,733		49.4
Votes counted	3,794,145		96.6	2,801,353		97.0
	V.C.	%	Seats	V.C.	%	Seats
Kommunistische Partei Österreichs (KPÖ) (extreme left/Communist)	176,656	0.5	0	20,497	0.7	0
Die Neutralen-Bürgerinitiative (left/pacifist)	48,600	1.3	0			
Die Grünen-Die Grüne Alternative (ecologist)	258,250	6.8	1	260,273	9.3	2
Sozialdemokratische Partei Österreichs (SPÖ) (left/Social Democrats)	1,105,910	29.1	6	888,338	31.7	7
Liberales Forum (LIF) (center/Liberal)	161,583	4.3	1	74,467	2.7	0
Österreichische Volkspartei (ÖVP) (right/Christian Democrat-Conservative)	1,124,921	29.6	7	859,175	30.7	7
Christlich Soziale Allianz (CSA) (right/Christian Democrat-Conservative (diss. ÖVP))				43,084	1.5	0
Freiheitliche Partei Österreichs (FPÖ) (extreme right/xenophobes)	1,044,604	27.5	6	65,519	23.4	5
Forum Handicap (miscellaneous)	32,621	0.9	0			
Total	3,794,145	100.0	21	2,801,353	100.0	21

Table A2.2 Belgium

Parties	1979 V.C.	%	Seats	1984 V.C.	%	Seats	1989 V.C.	%	Seats	1994 V.C.	%	Seats	1999 V.C.	%	Seats
Registered voters	6,800,584			6,975,677			7,096,273			7,211,311			7,343,466		
Voters	6,212,874	91.4		6,424,168	92.1		6,439,594	90.7		6,537,968	90.7		6,686,222	91.0	
Votes counted	5,442,978	87.6		5,721,894	89.1		5,899,285	91.6		5,966,755	91.3		6,223,166	93.1	
PVDA-PVDA-AE (extreme left/ Communist-Maoist)	36,602	1.1	0	30,555	0.9	0	20,746	0.6	0	41,816	1.1	0	21,998	0.6	0
RAL-SAP (extreme left/ Communist-Trotskyite)	10,702	0.3	0	14,910	0.4	0									
KPB (extreme left /Communist)	39,773	1.2	0	25,774	0.7	0									
REGEBO-RGB (extreme left)							26,471	0.7	0	15,549	0.4	0			
SO.LI.DE (left)													9,529	0.2	0
SP (left/Social Democrat)	698,889	20.9	3	979,702	28.1	4	733,247	20.0	3	651,371	17.6	3	550,237	14.2	2
ID21 (regionalist (Flanders) (diss. VU))	324,540	9.7	1	484,494	13.9	2	318,146	8.7	1	262,043	7.1	1	471,239	12.2	2
VOLKSUNIE (VU) (regionalist (Flanders))	34,706	1.0	0												
VVP (regionalist (Flanders) (diss. VU))															
AGALEV (ecologist)	77,986	2.3	0	246,712	7.1	1	446,524	12.2	1	396,198	10.7	1	464,043	12.0	2
CVP (center/Christian Democrat)	1,607,941	48.1	7	1,132,682	32.5	4	1,247,090	34.1	5	1,013,266	27.4	4	839,688	21.7	3
PVV-VLD (1992) (right/Liberal)	512,363	15.3	2	494,277	14.2	2	625,566	17.1	2	678,421	18.4	3	847,103	21.9	3
VTVANT (right/Liberal)													67,107	1.7	0
VLAAMS BLOK (VB) (extreme right/xenophobe-regionalist (Flanders) (diss. VU))				73,174	2.1	0	241,117	6.6	1	463,919	12.6	2	584,392	15.1	2
BEB (right/Conservative)										24,132	0.7	0			
PNPB (miscellaneous/populist)													17,095	0.4	0
WOW (miscellaneous /retired)										127,504	3.5	0			

	Votes	%	Seats	Votes	%	Seats	Votes	%	Seats	Votes	%	Seats	Votes	%	Seats
NWP (miscellaneous)										19,930	0.5	0			
Flemish-speaking subtotal	3,343,502	100.0	13	3,482,280	100.0	13	3,658,907	100.0	13	3,694,149	100.0	14	3,872,431	100.0	14
PTB-PTB-UA (extreme left/ Communist-Maoist)	8,821	0.4	0	13,082	0.6	0	9,029	0.4	0	17,454	0.8	0			
DEBOUT (extreme left)													46,089	2.0	0
LETD (extreme left /Communist-Trotskyite)							5,505	0.2	0	8,822	0.4	0			
LRT-POS (extreme-left / Communist-Trotskyite)	6,210	0.3	0	10,471	0.5	0	10,116	0.5	0						
GAUCHES UNIES (extreme left)										35,977	1.6	0			
EuropeNON (extreme left)	22,187	1.1	0												
PCB (extreme left /Communist)	106,023	5.0	0	61,605	2.8	0							25,541	1.1	0
PS (left/Social Democrat)	575,824	27.4	4	762,293	34.0	5	854,148	38.1	5	680,142	30.4	3	596,565	25.8	3
ECOLO-VERTS (ecologist)	107,833	5.1	0	220,663	9.9	1	371,053	16.6	2	290,859	13.0	1	525,313	22.7	3
SUD (center/Christian Democrat)										14,054	0.6	0			
PSC (center/Christian Democrat)	445,910	21.2	3	436,108	19.5	2	476,802	21.3	2	420,198	18.8	2	307,909	13.3	1
MCC (right/Christian Democrat (diss. PSC))															1
PRL (+FDF) (right/Liberal)	372,904	17.8	2	540,610	24.1	3	423,511	18.9	2	541,724	24.2	2	624,444	27.0	2
PLW (right/Liberal-regionalist (Wallonie))	17,566	0.8	0												
ERE (ecologist (diss. ECOLO-VERTS))									0						
FDF (regionalist (Wallonie))	414,603	19.7	1	142,879	6.4	0	85,870	3.8	0				cf. PRL		
FNFP-CFE (regionalist (Wallonie) (diss. RW))						0						1			
RASSEMBLEMENT WALLON (RW) (regionalist (Wallonie))			1	51,903	2.3	0									
VIVANT (right/Liberal)													55,135	2.4	0
FRONT NATIONAL (FN) (extreme right/xenophobe)										175,732	7.9	1	94,848	4.1	0

Table A2.2 (Continued)

Parties	1979 V.C.	%	Seats	1984 V.C.	%	Seats	1989 V.C.	%	Seats	1994 V.C.	%	Seats	1999 V.C.	%	Seats
FNB (extreme right/xenophobe (diss. FN))													24,792	1.1	0
AGIR (extreme right/ xenophobe-regionalist (Wallonie))										42,917	1.9	0			
PPB (extreme right/ecologists)	9,704	0.5	0										2,720	0.1	0
PCN (extreme right)	4,617	0.2	0												
POE (extreme right)													10,479	0.5	0
PNPB (miscellaneous/populist)	7,274	0.3	0												
PFU (miscellaneous/feminist)															
PH/HP (miscellaneous)							4,344	0.2	0	6,385	0.3	0			
French-speaking subtotal	*2,099,476*	*100.0*	*11*	*2,239,614*	*100.0*	*11*	*2,240,378*	*100.0*	*11*	*2,234,264*	*100.0*	*10*	*2,313,835*	*100.0*	*10*
PAB-PAB-AE (extreme left/ Communist-Maoist)										205	0.5	0	72	0.2	0
ECOLO (ecologist)										5,714	14.9	0	6,276	17.0	0
SP (left/Social Democrat)										4,820	12.6	0	4,215	11.4	0
CSP-EVP (center/Christian Democrat)										11,999	31.3	1	13,456	36.5	1
PDB (regionalist/Christian Democrat (German-speaking))										5,945	15.5	0	3,661	9.9	0
PFF (right/Liberal)										7,690	20.1	0	7,234	19.6	0
VIVANT (right/Liberal)													1,198	3.2	0
JUROPA (miscellaneous/students)										1,969	5.1	0	788	2.1	0
German-speaking subtotal										*38,342*	*100.0*	*1*	*36,900*	*100.0*	*1*
Total	*5,442,978*	*100.0*	*24*	*5,721,894*	*100.0*	*24*	*5,899,285*	*100.0*	*24*	*5,966,755*	*100.0*	*25*	*6,223,166*	*100.0*	*25*

Table A2.3 Denmark

Parties	1979			1984			1989			1994			1999		
Registered voters	3,754,423			3,878,600			3,923,549			3,994,200			4,009,594		
Voters	1,791,268	47.7		2,025,696	52.2		1,811,558	46.2		2,113,780	52.9		2,023,306	50.5	
Votes counted	1,754,211	97.9		2,001,906	98.8		1,789,395	98.8		2,079,937	98.4		1,970,276	97.4	
	V.C.	%	Seats	V.C.	%	Seats	V.C.	%	Seats	V.C.	%	Seats	V.C.	%	Seats
Danmarks Kommunistiske Parti (DKP) (extreme left/Communist	cf. FolkB														
Socialistisk Folkeparti (SF) (extreme left/post Communist)	81,991	4.7	1	183,580	9.2	1	162,902	9.1	1	178,543	8.6	1	140,053	7.1	1
Folkebevaegelsen mod EF/EU (miscellaneous/ anti-Europeans)	365,760	21.0	4	413,808	20.8	4	338,953	18.9	4	214,735	10.3	2	143,709	7.3	1
JuniBevaegelsen mod Unionen (miscellaneous/ anti-Europeans (diss. Folkebevaegelsen mod EU))										316,687	15.2	2	317,508	16.1	3
Danmarks Retsforbundet (left/Liberal)	60,964	3.5	0	cf. FolkB											
VenstreSocialisterne (VS) (left/Socialist)	59,379	3.4	0	25,305	1.3	0									
Socialdemokratiet (SD) (left/Social Democrat)	382,487	21.9	3	387,098	19.4	3	417,076	23.3	4	329,202	15.8	3	324,256	16.5	3
Det Radikale Venstre (RV) (left/Liberal)	56,944	3.3	0	62,560	3.1	0	50,196	2.8	0	176,480	8.5	1	180,089	9.1	1

Table A2.3 (Continued)

	1979			1984			1989			1994			1999		
	V.C.	%	Seats	V.C.	%	Seats	V.C.	%	Seats	V.C.	%	Seats	V.C.	%	Seats
Kristeligt Folkeparti (KRF) (center/Christian Democrat-Conservative Protestant)	30,985	1.8	0	54,624	2.7	0	47,768	2.7	0	22,986	1.1	0	39,128	2.0	0
Centrum–Demokraterne (CD) (center/Social Democrat)	107,790	6.2	1	131,984	6.6	1	142,190	7.9	2	18,365	0.9	0	68,717	3.5	0
Venstre–Danmarks Liberale Parti (LV) (right/Liberal)	252,767	14.5	3	248,397	12.5	2	297,565	16.6	3	394,362	19.0	4	460,834	23.4	5
Det Konservative Folkeparti (KF) (right/Conservative)	245,309	14.1	2	414,177	20.8	4	238,760	13.3	2	368,890	17.7	3	166,884	8.5	1
Dansk Folkeparti (DFP) (extreme right/populist (diss. FRP))													114,865	5.8	1
Fremskridtspartiet (FRP) (extreme right/populist)	100,702	5.8	1	68,747	3.5	0	93,985	5.3	0	59,687	2.9	0	14,233	0.7	0
Total	*1,745,078*	*100.0*	*15*	*1,990,280*	*100.0*	*15*	*1,789,395*	*100.0*	*16*	*2,079,937*	*100.0*	*16*	*1,970,276*	*100.0*	*16*
Siumut (regionalist/Social Democrat (Greenland))	5,053	55.3	1	7,386	63.5	1									
Atassut (regionalist/Liberal (Greenland))	4,080	44.7	0	4,240	36.5	0									
Greenland subtotal	*9,133*	*100.0*	*1*	*11,626*	*100.0*	*1*									
Total[1]	*1,754,211*	*100.0*	*16*	*2,001,906*	*100.0*	*16*									

[1]Greenland has not been a member of the European Community since December 31, 1984.

Table A2.4 Finland

Parties	1996			1999		
Registered voters	4,108,703			4,141,098		
Voters	2,366,504		57.6	1,248,122		30.1
Votes counted	2,249,411		95.1	1,242,303		99.5
	V.C.	%	Seats	V.C.	%	Seats
Suomen Kommunistinen Puolue (SKP) (extreme left/Communist)				7,556	0.6	0
Vasemmistoliitto (VAS) (extreme left/ Post-Communist)	236,490	10.5	2	112,757	9.1	1
Suomen Sosialdemokraattinen Puolue (SDP) (left/ Social Democrat)	482,577	21.5	4	221,836	17.9	3
Vihreä Liitto (VIHR) (ecologist)	170,670	7.6	1	166,786	13.4	2
Liberaalinen Kansanpuolue (LKP) (center/Liberal)	8,305	0.4	0			
Suomen Keskusta (KESK) (center/agrarians)	548,041	24.4	4	264,640	21.3	4
Svenska Folkpartiet i Finland (SFP-RKP) (center/regionalist (Swedish-speakers))	129,425	5.8	1	84,153	6.8	1
Suomen Kristillinen Liitto (SKL) (center/Christian Democrat-Conservative Protestant)	63,279	2.8	0	29,637	2.4	1
Kansallinen Kokoomus (KOK) (right/Conservative)	453,729	20.2	4	313,960	25.3	4
Nuorsuomalainen Puolue (NUORS) (right/Liberal)	68,134	3.0	0			
Perussuomalaiset (PERUS) (right/agrarians-Conservative Protestant)	15,004	0.7	0	9,854	0.8	0
Kirjava Puolue (KIPU) (ecologist)				29,215	2.4	0
Eläkeläiset Kansan Asialla (EKA) (miscellaneous/retired)	2,640	0.1	0	1,909	0.2	0
Suomen Eläkeläisten Puolue (SEP) (miscellaneous/retired)	6,357	0.3	0			
Vapaan Suomen Liitto (VSL) (extreme right/populist)	13,746	0.6	0			
UE Alternative (miscellaneous/ anti-Europeans)	47,687	2.1	0			
Luonnonlain Puolue (LLP) (miscellaneous)	3,327	0.1	0			
Total	2,249,411	100.0	16	1,242,303	100.0	16

Table A2.5 France

Parties	1979			1984			1989			1994			1999		
Registered voters	35,180,531			36,880,688			37,633,138			37,724,297			40,132,517		
Voters	21,356,960			20,918,772			18,572,672			20,276,895			18,766,155		
Votes counted	20,242,347		60.7	20,180,934		56.7	18,040,332		49.3	19,218,608		53.7	17,647,172		46.8
	V.C.	%	Seats	V.C.	%	Seats	V.C.	%	Seats	V.C.	%	Seats	V.C.	%	Seats
Lutte ouvrière (extreme left/Trotskyite)	623,663	3.1		417,702	2.1		256,865	1.4		442,723	2.3		914,811	5.2	5
MPPT (extreme left/Trotskyite)				182,320	0.9		109,465	0.6		84,513	0.4				
PCF (Communist left)	4,153,710	20.5	19	2,261,312	11.2	10	1,396,486	7.7	7	1,342,222	6.7	7	1,196,491	6.8	6
PSU (miscellaneous left)	332	0		146,238	0.7										
Rénovateurs Communist (miscellaneous left)							74,256	0.4							
MDC (miscellaneous left)										494,986	2.5				
PS (Socialist left)	4,763,026	23.5	22	4,188,875	20.7	20	4,252,828	23.6	22	2,824,173	14.5	15	3,874,231	21.9	22
Radical (center left)				670,474	3.3					2,344,457	12	13			
Verts (ecologist)	888,134	4.4		680,080	3.4		1,915,620	10.6	9	574,806	3	2	1,715,729	9.7	9
Génération Ecologie (ecologist)										392,291	2				
MEI (ecologist)													268,038	1.5	
Hallier (regionalist)	337	0								76,436	0.4				
Siméoni (regionalist)										37,041	0.2				
Moutoussamy (regionalist)															
Jos (regionalist)													1,707	0	
Cartan (federalist)				78,234	0.4										
Biancheri (federalist)							31,459	0.2							
Touati A. (federalist)										71,814	0.4				

Allenbach (federalist)														0	0
Servan-Schreiber (center right)	373,259	1.8													
Gomez (center right)															
Veil (center right)	382,404	1.9					1,517,653	8.4	7						
UDF (center right and Liberal)	5,588,851	27.6	25										1,638,999	9.3	9
RPR (Gaullist right)	3,301,980	16.3	15	8,683,596	43	41	5,200,289	28.8	26				2,263,201	12.8	12
RPR-UDF (united right)										4,985,574	25.6	28			
CNIP (miscellaneous right)	283,144	1.4		123,642	0.6		58,639	0.3							
Touati G. (miscellaneous right)				138,220	0.7					125,340	0.6				
Nicoud (miscellaneous right)															
Joyeux (miscellaneous right)							136,014	0.8							
De Villiers-MPF (miscellaneous right)										2,404,105	12.3	13	2,304,544	13	13
Aillaud (miscellaneous right)										290	0				
RPF (miscellaneous right sovereignist)	265,911	1.3													
Maudrux (miscellaneous right)													124,561	0.7	
Miguet (miscellaneous right)													312,450	1.8	
PFN (extreme right)															
FN (extreme right)				2,210,334	10.9	10	2,121,836	11.8	10	2,050,086	10.5	11	1,005,285	5.7	5
POE (extreme right)				17,503	0		32,169	0.2							
MN (extreme right)													578,837	3.3	
LN (extreme right)													683	0	
Alessandri (miscellaneous)							187,346	1							
Schwartzenberg (miscellaneous)										305,633	1.6				
Cotten (miscellaneous)										56,668	0.3		274	0	
PLN (miscellaneous)										103,261	0.5		71,409	0.4	
Larrourou (miscellaneous)													178,064	1	
Chanut-Sapin (miscellaneous)													1,995	0	
CPNT (miscellaneous/hunters)							749,407	4.1		771,061	4		1,195,863	6.8	6
Total			81			81			81			87			87

Table A2.6 Germany

Parties	1979			1984			1989			1994			1999		
	V.C.	%	Seats	V.C.	%	Seats	V.C.	%	Seats	V.C.	%	Seats	V.C.	%	Seats
Registered voters	42,751,940			44,465,989			45,773,179			60,473,927			60,786,904		
Voters	28,098,872	65.7		25,238,754	56.8		28,508,598	62.3		36,295,529	60.0		27,468,932	45.2	
Votes counted	27,847,109	99.1		24,851,371	98.5		28,206,690	98.9		35,411,414	97.6		27,059,273	98.5	
MLPD (extreme left/ Communist-Maoist)							10,134	0.0	0						
PEAD (extreme left/ Communist-Trotskyite)							10,377	0.0	0	12,992	0.0	0			
BSA (extreme left/ Communist-Trotskyite)							7,788	0.0	0	10,678	0.0	0			
PDS (extreme left/ Communist)										1,670,316	4.7	0	1,567,745	5.8	6
DKP (extreme left/ Communist)	112,055	0.4	0				57,704	0.2	0						
DIE FRIEDENSLISTE (extreme left/Pacifist)				313,108	1.3	0									
AUTONOME (extreme left)										37,672	0.1	0			
NEUES FORUM (left/pacifist)										107,615	0.3	0			
BÜNDNIS 90 (left/pacifist)															

Party	Votes	%	Seats	Votes	%	Seats	Votes	%	Seats	Votes	%	Seats	Votes	%	Seats
DIE GRÜNEN (ecologist)	893,683	3.2	0	2,025,972	8.2	7	2,382,102	8.4	8	3,563,268	10.1	12	1,741,494	6.4	7
ÖDP (ecologist)				77,026	0.3	0	184,309	0.7	0	273,776	0.8	0	100,048	0.4	0
ÖKO-UNION (ecologist)							55,463	0.2	0						
DIE TIERSCHUTZPARTEI (ecologist)													185,186	0.7	0
SPD (left/Social Democrat)	11,370,045	40.8	35	9,295,417	37.4	33	10,525,728	37.3	31	11,389,697	32.2	40	8,307,085	30.7	33
FDP (center/Liberal)	1,662,621	6.0	4	1,192,624	4.8	0	1,576,715	5.6	4	1,442,857	4.1	0	820,371	3.0	0
ZENTRUM (center/Christian Democrat)	31,367	0.1	0	93,921	0.4	0	41,190	0.1	0	110,778	0.3	0	7,080	0.0	0
BAYERNPARTEI (BP) (regionalist/Christian Democrat (Bavaria))				23,539	0.1	0	71,991	0.3	0				14,950	0.1	0
CDU (right/Christian Democrat-Conservative)	10,883,085	39.1	34	9,308,411	37.5	34	8,332,846	29.5	25	11,346,073	32.0	39	10,628,224	39.3	43
CSU (right/Christian Democrat-Conservative-regionalist (Bavaria))	2,817,120	10.1	8	2,109,130	8.5	7	2,326,277	8.2	7	2,393,374	6.8	8	2,540,007	9.4	10
CBV (right/Christian Democrat-Conservative-regionalist (Bavaria))	45,311	0.2	0												
DSU (right/Christian Democrat-Conservative-regionalist (ex CDR))										80,618	0.2	0			
BUND FREIER BÜRGER (BfB) (extreme right/populist (diss. FDP))										385,676	1.1	0			

Table A2.6 (Continued)

Parties	1979			1984			1989			1994			1999		
	V.C.	%	Seats	V.C.	%	Seats	V.C.	%	Seats	V.C.	%	Seats	V.C.	%	Seats
CHRISTLICHE MITTE (CM) (extreme right/ Fundamentalist (diss. CDU))							43,580	0.2	0	66,766	0.2	0	30,746	0.1	0
CHRISTLICHE LIGA (extreme right/ Fundamentalist)							30,879	0.1	0	40,115	0.1	0			
PBC (extreme right/ Fundamentalist)										93,210	0.3	0	68,732	0.3	0
DIE REPUBLIKANER (REP) (extreme right/ xenophobe)							2,008,629	7.1	6	1,387,070	3.9	0	461,038	1.7	0
DVU (extreme right/ xenophobe)									0						
NPD (extreme right/xenophobe)				198,633	0.8	0	444,921	1.6	0	77,227	0.2	0	107,662	0.4	0
PATRIOTEN (extreme right/xenophobe)							12,907	0.0	0						
FAP (extreme right/ xenophobe)							19,151	0.1	0						
EAP-BüSolidarität (BüSo) (extreme right)	31,822	0.1	0	30,874	0.1	0	32,246	0.1	0	23,851	0.1	0	9,431	0.0	0
MÜNDIGE BÜRGER (miscellaneous/populist)				52,753	0.2	0									
STATT PARTEI (miscellaneous/ populist (diss. CDU))										168,738	0.5	0			

EFP (miscellaneous/ Eurofederalists»	34,500	0.1	0												
FRAUENPARTEI (miscellaneous Feminist)	94,463	0.4	0												
DIE FRAUEN miscellaneous/Feminist													100,128	0.4	0
FAMILIE (miscellaneous/ Families-regionalist (Sarre))										2,781	0.0	0	4,117	0.0	0
DIE GRAUEN (miscellaneous/retired)										275,866	0.8	0	112,142	0.4	0
PASS (miscellaneous/ Unemployed)										127,104	0.4	0	71,430	0.3	0
APD (miscellaneous/ Automobilists)										231,265	0.7	0	97,984	0.4	0
ASP (miscellaneous/ Automobilists)													34,029	0.1	0
NATURGESETZ PARTEI (miscellaneous)										92,031	0.3	0	38,139	0.1	0
HUMANISTISCHE PARTEI (HP) (miscellaneous)							10,885	0.0	0				11,505	0.0	0
BEWUSSTSEIN (miscellaneous)							20,868	0.1	0						
Total	*27,847,109*	*100.0*	*81*	*24,851,371*	*100.0*	*81*	*28,206,690*	*100.0*	*81*	*35,411,414*	*100.0*	*99*	*27,059,273*	*100.0*	*99*

Table A2.7 Great Britain

Parties	1979 V.C.	1979 %	1979 Seats	1984 V.C.	1984 %	1984 Seats	1989 V.C.	1989 %	1989 Seats	1994 V.C.	1994 %	1994 Seats	1999 V.C.	1999 %	1999 Seats
Registered voters	41,155,166			42,493,274			43,180,720			43,443,944			44,126,427		
Voters	13,494,324	32.8		14,034,958	33.0		n.c.			n.c.			10,689,846	24.2	
Votes counted	13,446,091	99.6		13,998,190	99.7		15,896,078	36.8		15,827,417	36.4		10,682,080	99.9	
Scottish National Party (SNP) (regionalist)	247,836	1.9	1	230,594	1.7	1	406,686	2.6	1	487,239	3.1	2	268,528	2.7	2
Plaid Cymru (PC) (regionalist (Wales))	83,399	0.6	0	103,031	0.8	0	115,062	0.7	0	162,478	1.0	0	185,235	1.9	2
Ecology Party (EP)–Green Party (GP) (1985)	17,953	0.1	0	70,853	0.5	0	2,292,718	14.9	0	494,561	3.1	0	625,378	6.3	2
Socialist Party of GB (SPGB) (extreme left)							919	0.0	0				1,510	0.0	0
Scottish Socialist Party (SSP) (left/labor-regionalist)													39,720	0.4	0
Socialist Labour Party (SLP) (left/labor)													86,749	0.9	0
Independent-MEP Independent Labour (left/labor)													36,849	0.4	0
Alternative Labour List (ALAB) (left/labor)													26,963	0.3	0
Labour Party (LAB) (+Co-operative Party) (left/labor)	4,253,207	33.0	17	4,865,261	36.5	32	6,153,661	40.1	45	6,753,860	42.7	62	2,803,821	28.0	29
Social Democratic Party (old SDP) (center/labor)							75,886	0.5	0						
Social Democratic Party (SDP) (center/labor)						0									
Liberal Party (LIB) (center/Liberal)	1,690,599	13.1	0	2,591,635	ALL 19.5	0	944,861	6.2	0						
Social and Liberal Democrats (LD) (1988/89) (center/Liberal)										2,552,730	16.1	2	1,266,549	12.7	10

Party	Votes	%	Seats	Votes	%	Seats	Votes	%	Seats	Votes	%	Seats	Votes	%	Seats
Liberal Party (center/Liberal)													93,051	0.9	0
Pro Euro Conservative Party (PECP) (right/Conservative)													138,097	1.4	0
Conservative and Unionist Party (CP) (right/Conservative)	6,508,493	50.6	60	5,426,821	40.8	45	5,331,098	34.7	32	4248531	26.8	18	3,578,217	35.8	36
UK Independence Party (UKIP) (right/Nationalist)													696,057	7.0	3
British National Party (BNP) (extreme right/xenophobe)													102,647	1.0	0
Natural Law Party (NLP) (miscellaneous)													21,327	0.2	0
Other candidates	72,365	0.6	0	24,678	0.2	0	40,376	0.3	0				32,573	0.3	0
Great Britain subtotal	1.3E+07	100.0	78	13,312,873	100.0	78	15,361,267	100.0	78	n.c.		84	10,003,271	100.0	84
Sinn Féin (SF) (regionalist/Nationalist (Ulster))				91,476	13.3	0	48,914	9.1	0				117,643	17.3	0
Workers Party (WP) (extreme left/Nationalist-regionalist (Ulster))	4,418	0.8	0	8,712	1.3	0	5,590	1.0	0						
NI Ecology Party (NIEP)–Green Party (GP)				2,172	0.3	0	6,569	1.2	0						
Social Democratic & Labour Party (SDLP) (left/nationalist-labor-regionalist (Ulster))	140,622	24.6	1	151,399	22.1	1	136,335	25.5	1	161992	1.0	1	190,731	28.1	1
Alliance Party of NI (APNI) (center/Pacifist)	39,026	6.8	0	34,046	5.0	0	27,905	5.2	0				14,391	2.1	0
Progressive Unionist Party of NI (PUP) (right/Conservative Protestant)													22,494	3.3	0
Unionist Party of NI (UPNI) (right/Conservative Protestant)	3,712	0.6	0												
Conservative Party in Northern Ireland (right/Conservatve Protestant)							25,789	4.8	0						

Table A2.7 (Continued)

Parties	1979			1984			1989			1994			1999		
	V.C.	%	Seats	V.C.	%	Seats	V.C.	%	Seats	V.C.	%	Seats	V.C.	%	Seats
UK Unionist Party (UKUP) (right/Conservative Protestant)													20,283	3.0	0
Ulster Unionist Party (UUP) (right/ Conservative Protestant)	125,169	21.9	1	147,169	21.5	1	118,785	22.2	1	133,459	0.8	1	119,507	17.6	1
Ulster Popular Unionist Party (UPUP) (right/Conservative Protestant)	38,198	6.7	0	20,092	2.9	0									
Ulster Democratic Unionist Party (DUP) (extreme right/Conservative Protestant)	170,688	29.8	1	230,251	33.6	1	160,110	29.9	1	163,246	1.0	1	192,762	28.4	1
Natural Law Party (NLP) (miscellaneous)													998	0.1	0
Other candidates	50,406	8.8	0				4,814	0.9	0						
North Ireland subtotal	572,239	100.0	3	685,317	100.0	3	534,811	100.0	3	n.c		3	678,809	100.0	3
Other candidates										669,321	4.2	0			
Total	1.3E + 07	100.0	81	13,998,190	100.0	81	15,896,078	100.0	81	15,827,417		87	10,682,080	100.0	87

Table A2.8 Greece

Parties	1981			1984			1989			1994			1999		
	V.C.	%	Seats	V.C.	%	Seats	V.C.	%	Seats	V.C.	%	Seats	V.C.	%	Seats
Registered voters	7,059,778			7,760,669			8,347,387			9,550,596			9,387,417		
Votes	5,752,344	81.5		6,012,203	80.6		6,668,113	79.9		6,803,884	71.2		6,711,728	71.5	
Votes counted	5,677,661	98.7		5,956,588	99.1		6,544,669	98.1		6,532,690	96.0		6,427,648	95.8	
KKE (extreme left/ Communist)	729,052	12.8	3	693,466	11.6	3		SYN	3	410,747	6.3	2	557,134	8.7	3
KKE-es-EAR-SYN (extreme left/ Post-Communist)	300,841	5.3	1	203,671	3.4	1	936,175	14.3	1	408,072	6.2	2	332,116	5.2	2
KO.DI.SO (extreme left)	241,666	4.3	1	47,400	0.8	0			0						
KAE (extreme left)			0												
EDA (extreme left)			1			1			0						
DI.K.KI (left/Socialist-nationalist (diss. PASOK))													440,079	6.8	2
PA.SO.K (left/Socialist)	2,278,030	40.1	8	2,477,445	41.6	7	2,352,271	35.9	9	2,458,657	37.6	10	2,116,507	32.9	9
PAR.KE (center)			1			2									
E.DI.K (center)	63,673	1.1	0	16,865	0.3	0	18,313	0.3	0						
CHRISTIANIK	65,056	1.1	0	26,731	0.4	0	26,970	0.4	0						
DIMOKRATIA (center/ Christian Democrat)							26,040	0.4	0						
KOMMA FILELEFTHERON (center/Liberal)	59,141	1.0	0	20,910	0.4	0									
ENOSI KENTROON (center)										77,952	1.2	0	52,457	0.8	0
I FILELEFTHEI-S. MANOS (right+Liberal Conservative (diss. ND))													103,850	1.6	0

Table A2.8 (Continued)

Parties	1981			1984			1989			1994			1999		
	V.C.	%	Seats	V.C.	%	Seats	V.C.	%	Seats	V.C.	%	Seats	V.C.	%	Seats
DI.ANA (right/Conservative (diss. ND))							89,469	1.4	1	182,525	2.8	0			
NEA DIMOKRATIA (ND) (right/Conservative)	1,779,462	31.3	8	2,266,088	38.0	9	2,647,215	40.4	10	2,133,405	32.7	9	2,313,265	36.0	9
POL.AN (right/Conservative-nationalist (diss. ND))										564,787	8.6	2	146,760	2.3	0
KOMMA PROODEFTIKON (KP) (extreme right)	111,245	2.0	1	10,136	0.2	0									
KEME (extreme right)	49,495	0.9	0												
E.PEN (extreme right)				136,623	2.3	1	75,877	1.2	0	50,749	0.8	0	48,726	0.8	0
CHRYSI AVGHI (extreme right)										7,264	0.1	0			
KOMMA ELLINON KYNIGON (miscellaneous/hunters)										41,158	0.6	0	64,063	1.0	0
OIK.ENAL (ecologist)							72,369	1.1	0	17,017	0.3	0	14,822	0.2	0
KOLLATOS (ecologist)							67,998	1.0	0	45,871	0.7	0	42,539	0.7	0
Other lists				57,253	1.0	0	231,972	3.5	0	134,486	2.1	0	195,330	3.0	0
Total	5,677,661	100.0	24	5,956,588	100.0	24	6,544,669	100.0	24	6,532,690	100.0	25	6,427,648	100.0	25

Table A2.9 Ireland

Parties	1979			1984			1989			1994			1999		
	V.C.	%	Seats	V.C.	%	Seats	V.C.	%	Seats	V.C.	%	Seats	V.C.	%	Seats
Registered voters	2,188,798			2,413,404			2,453,451			2,631,575			2,864,361		
Votes	1,392,285	63.6		1,147,745	47.6		1,675,119	68.3		1,157,296	44.0		1,438,287	50.2	
Votes counted	1,339,072	96.2		1,120,416	97.6		1,632,728	97.5		1,137,490	98.3		1,391,740	96.8	
Sinn Féin (SF) (nationalist)	43,942	3.3	0	54,672	4.9	0	37,127	2.3	0	33,823	3.0	0	88,165	6.3	0
SF-WP—Workers Party of Ireland (1982) (extreme left/Communist-Trotskyites)				48,449	4.3	0	123,265	7.5	1	22,100	1.9	0			
Democratic left (DL) (extreme left (diss. WP))										39,706	3.5	0			
Irish Labour Party (ILP) (left/labor)	193,898	14.5	4	93,656	8.4	0	155,782	9.5	1	124,972	11.0	1	121,542	8.7	1
Green Alliance (GA)-Green Party (GP) (1988) (ecologist)				5,242	0.5	0	61,041	3.7	0	90,046	7.9	2	93,100	6.7	2
Fine Gael (FG) (center/Liberal)	443,652	33.1	4	361,034	32.2	6	353,094	21.6	4	276,095	24.3	4	342,171	24.6	4
Progressive Democrats (PD) (center/nationalist-Liberal (diss. FF))							194,059	11.9	1	73,696	6.5	0			
Fianna Fáil (FF) (right/nationalist)	464,451	34.7	5	438,946	39.2	8	514,537	31.5	6	398,066	35.0	7	537,757	38.6	6
Other candidates	193,129	14.4	2	118,417	10.6	1	193,823	11.9	2	78,986	6.9	1	209,005	15.0	2
Total	1,339,072	100.0	15	1,120,416	100.0	15	1,632,728	100.0	15	1,137,490	100.0	15	1,391,740	100.0	15

Table A2.10 Italy

Parties	1979 V.C.	%	Seats	1984 V.C.	%	Seats	1989 V.C.	%	Seats	1994 V.C.	%	Seats	1999 V.C.	%	Seats
Registered voters	**42,203,405**			**44,400,652**			**46,346,961**			**48,461,792**			**49,278,309**		
Voters	36,148,180	85.7		37,096,082	83.5		37,572,759	81.1		35,667,440	73.6		34,376,459	69.8	
Votes counted	35,042,601	96.9		35,141,553	94.7		34,628,023	92.2		32,949,725	92.4		31,110,065	90.5	
COBAS PER L'AUTOR-GANIZZAZIONE (extreme left)													4,370	0.0	0
DEMOCRAZIA PROLETARIA (DP) (extreme left/Communist)	252,342	0.7	1	506,753	1.4	1	449,639	1.3	1						
PDUP (extreme left/Communist)	406,656	1.2	1												
PCI-RC (1991) (extreme left/Communist)	10,361,344	29.6	24	11,714,428	33.3	26	9,598,369	27.6	22	2,004,716	6.1	5	1,328,515	4.3	4
COMUNISTI ITALIANI (CI) (extreme left/Communist (diss. RC))													622,259	2.0	2
PDS (ex PCI)-DS (1998) (left)	3,866,946	11.0	9	3,940,445	11.2	9	5,151,929	14.8	12	6,281,354	19.1	16	5,395,363	17.3	15
PSI - PSI-AD (1994)—SDI (left/Socialist)										606,538	1.8	2	671,821	2.2	2
PARTITO SOCIALISTA (left/Socialist)													42,554	0.1	0
PSDI (left/Social Democrat)	1,514,272	4.3	4	1,225,462	3.5	3	945,383	2.7	2	227,439	0.7	1			
VERDI ARCOBALENO (ecologist)							830,980	2.4	2						
FEDERAZIONE DEI VERDI (ecologist)							1,317,119	3.8	3	1,055,797	3.2	3	548,908	1.8	2
RETE (left/Christian Democrat)										366,258	1.1	1			

Party	Votes	%	Seats	Votes	%	Seats	Votes	%	Seats	Votes	%	Seats	Votes	%	Seats
I DEMOCRAT-CI-PRODI (left/Christian Democrat)													2,407,952	7.7	7
SÜDTIROLER VOLKSPARTEI (SVP) (center/Christian Democrat-regionalist (Haut Adige))	196,373	0.6	1	198,220	0.6	1	172,383	0.5	1	202,668	0.6	1	155,751	0.5	(1)
RINNOVAMENTO ITALIANO-DINI (RI) (center/Christian Democrat)													353,806	1.1	1
U.D.EUR (center/Christian Democrat)													499,498	1.6	1
DC-PPI (1994) (center/Christian Democrat)	12,774,320	36.5	29	11,583,767	33.0	26	11,451,053	32.9	26	3,295,337	10.0	8	1,319,499	4.2	4
CDU (center/Christian Democrat (diss. PPI))													670,065	2.2	2
CCD (right/Christian Democrat (diss. DC))													806,429	2.6	2
FORZA ITALIA (FI) (right/Liberal)										10,089,139	30.6	27	7,829,624	25.2	22
PLI-LIBERALI (right/Liberal)	1,271,159	3.6	3	2,140,501	6.1	2	POLO		0	53,983	0.2	0	168,178	0.5	1
PRI (center/Liberal)	896,139	2.6	2				LAICO		3	242,786	0.7	1	2,631,205	8.5	7
PR-RIF-PANNELLA-BONINO (left/Liberal)	1,285,065	3.7	3	1,199,876	3.4	3	1,532,388	4.4	1	702,717	2.1	2			
L.LOMBARDA (LL)-L.NORD (LN) (1991) (right/regionalist (Lombardy))							636,242	1.8	2	2,162,586	6.6	6	1,395,547	4.5	4
LEGA VENETA (regionalist (Venetia))	164,115	0.5	0										118,104	0.4	0
LEGA ALPINA LUMBARDA (regionalist (Lombardy) (diss. Lega Nord))										110,458	0.3	0			
PATTO SEGNI (center/Christian Democrat (diss. DC))										1,073,095	3.3	3			

Table A2.10 (Continued)

Parties	1979			1984			1989			1994			1999		
	V.C.	%	Seats	V.C.	%	Seats	V.C.	%	Seats	V.C.	%	Seats	V.C.	%	Seats
ALLEANZA NAZIONALE (AN) (right/Nationalist)										4,108,670	12.5	11	3,202,895	10.3	9
MSI-DN-MSFT (extreme right)	1,909,055	5.4	4	2,274,556	6.5	5	1,918,650	5.5	4			(1)	495,351	1.6	1
DEMOCRAZIA NAZIONALE (DN-DC) (extreme right)	142,537	0.4	0												
LEGA d'AZIONE MERIDIONALE-CITO (regionalist (Pouilles))				FED		0	FED		0	224,033	0.7	0	93,353	0.3	0
UNION VALDÔTAINE (UV) (regionalist (Val d'Aosta))	166,393	0.5	0	193,430	0.6	1	207,739	0.6	1	126,937	0.4	0	41,227	0.1	0
PARTITO SARDO d'AZIONE (PSdA) ... (regionalist (Sardinia))							430,150	1.2	1				60,534	0.2	0
LEGA ANTIPROIBIZIONISTI DROGA (miscellaneous/antiprohibitionist)															
PARTITO PENSIONATI (miscellaneous/retired)							162,293	0.5	0	15,214	0.0	0	232,169	0.7	1
PARTITO UMANISTA (miscellaneous)													15,088	0.0	0
Total	3.5E+07	100.0	81	35,141,553	100.0	81	34,804,317	100.0	81	32,949,725	100.0	87	31,110,065	100.0	87

Table A2.11 Luxembourg

Parties	1979			1984			1989			1994			1999		
	V.C.	%	Seats	V.C.	%	Seats	V.C.	%	Seats	V.C.	%	Seats	V.C.	%	Seats
Registered voters	**212,740**			**215,792**			**218,940**			**224,031**			**228,712**		
Voters	**189,141**	**88.9**		**191,602**	**88.8**		**191,342**	**87.4**		**198,370**	**88.5**		**199,597**	**87.3**	
Votes counted	**170,759**	**90.3**		**173,888**	**90.8**		**174,472**	**91.2**		**178,643**	**90.1**		**180,839**	**90.6**	
LCR—RSP (extreme left/Communist–Trotskyite)	5,085	0.5	0	3,791	0.4	0	6,053	0.6	0						
PCL-KPL (extreme left/Communist)	48,813	5.0	0	40,395	4.1	0	46,791	4.7	0	16,559	1.6	0			
Déi Lénk (extreme left/post-Communist)										9,421	0.9	0	28,226	2.8	0
Gréng Alternativ Allianz (GRAL) (extreme left/ecologists (diss. GAP))							8,577	0.9	0						
Alternativ Lëscht (AL)—GAP (1983) (extreme left/ecologists)	9,845	1.0	0	60,152	6.1	0	42,926	4.3	0						
GLEI (ecologist)							61,054	6.1	0						
Déi Gréng (1993) (ecologist)										110,888	10.9	1	108,693	10.7	1
Gréng a Liberal Allianz (GAL) (ecologist (diss. Déi Gréng))													18,631	1.8	0
PSI (left/Social Democrat (diss. LSAP-POSL))				25,355	2.6	0									

Table A2.11 (Continued)

Parties	1979			1984			1989			1994			1999		
	V.C.	%	Seats	V.C.	%	Seats	V.C.	%	Seats	V.C.	%	Seats	V.C.	%	Seats
LSAP-POSL (left/Social Democrat	211,106	21.6	1	296,382	29.9	2	252,920	25.4	2	251,500	24.8	2	239,502	23.6	2
PSD-SDP (left/Social Democrat)	68,289	7.0	0												
CSV-PCS (center/Christian Democrat)	352,296	36.1	3	345,586	34.9	3	346,621	34.9	3	319,462	31.5	2	321,471	31.7	2
NOMP (right/Christian Democrat (diss. CVP-PSC))										8,765	0.9	0			
DP-PD (right/Liberal)	274,307	28.1	2	218,481	22.1	1	198,254	19.9	1	190,977	18.8	1	207,665	20.5	1
PL-LP (right/Liberal)	5,610	0.6	0												
GLS (right/Christian Democrat-nationalist (diss. CVP-PSC))										12,091	1.2	0			
National Bewegong (extreme right/xenophobe)							28,867	2.9	0	24,141	2.4	0			
ADR (miscellaneous/retired-populist)										70,470	6.9	0	91,253	9.0	0
Firwat nët?/Why not? (miscellaneous)							1,876	0.2	0						
Total (1)	975,351	100.0	6	990,142	100.0	6	993,939	100.0	6	1,014,274	100.0	6	1,015,441	100.0	6

Table A2.12 Netherlands

Parties	1979			1984			1989			1994			1999		
	V.C.	%	Seats	V.C.	%	Seats	V.C.	%	Seats	V.C.	%	Seats	V.C.	%	Seats
Registered voters	9,808,176			10,485,014			11,099,123			11,618,677			11,862,864		
Voters	5,700,603	58.1		5,334,582	50.9		5,270,374	47.5		4,146,730	35.7		3,560,764	30.0	
Votes counted	5,667,303	99.4		5,296,749	99.3		5,242,333	99.5		4,133,557	99.7		3,544,408	99.5	
Socialistische Partij (SP) (extreme left)							34,332	0.7	0	55,311	1.3	0	178,642	5.0	1
EVP (extreme left/Christian Democrat)									0						
PPR (extreme left/Christian Democrat)	92,055	1.6	0			1			1						
CPN (extreme left/Communist)	97,343	1.7	0			0									
PSP (extreme left/Pacifist)	97,243	1.7	0		GPA	1									
GPN (extreme left/ecologist)				296,488	5.6	0			0						
Regenboog-Groenlinks (GL) (1990) (extreme left/ecologist)							365,535	7.0		154,547	3.7	1	419,869	11.8	4
De Groenen (ecologist)				67,413	1.3	0				97,206	2.4	0			
Lijst de Groenen (ecologist)										8,844	0.2	0			
PvdA (left/Social Democrat)	1,722,240	30.4	9	1,785,165	33.7	9	1,609,626	30.7	8	945,869	22.9	8	712,929	20.1	6
Democraten 66 (D66) (left/Liberal)	511,967	9.0	2	120,826	2.3	0	311,990	6.0	1	481,843	11.7	4	205,623	5.8	2
CDA (center/Christian Democrat)	2,017,743	35.6	10	1,590,218	30.0	8	1,814,107	34.6	10	1,271,855	30.8	10	954,898	26.9	9
VVD (right/Liberal)	914,787	16.1	4	1,002,685	18.9	5	714,745	13.6	3	740,443	17.9	6	698,050	19.7	6
RPF (right/Conservative Protestant)						0			0			0			1
GPV (right/Conservative Protestant)	62,610	1.1	0			0			0			1			1

Table A2.12 (Continued)

Parties	1979			1984			1989			1994			1999		
	V.C.	%	Seats	V.C.	%	Seats	V.C.	%	Seats	V.C.	%	Seats	V.C.	%	Seats
SGP (right/Conservative Protestant)	126,412	2.2	0	275,786	5.2	1	309,060	5.9	1	322,793	7.8	1	309,612	8.7	1
CentrumDemocraten (CD) (extreme right/xenophobe (diss. CP))							40,780	0.8	0	43,299	1.0	0	17,740	0.5	0
Centrum Partij (CP) (extreme right/xenophobe)				134,877	2.5	0									
De Europese Partij (DEP) (extreme right)							23,298	0.4	0				23,231	0.7	0
IDE (miscellaneous/ Eurofederalists)	24,903	0.4	0										13,234	0.4	0
EVPN (miscellaneous/retired)							18,860	0.4	0						
God Met Ons (miscellaneous)				23,291	0.4	0									
Lijst Leschot (miscellaneous)										11,547	0.3	0			
Een Betere Toekomst (miscellaneous)													10,580	0.3	0
Lijst Sala (miscellaneous)															
Total	*5,667,303*	*100.0*	*25*	*5,296,749*	*100.0*	*25*	*5,242,333*	*100.0*	*25*	*4,133,557*	*100.0*	*31*	*3,544,408*	*100.0*	*31*

Table A2.13 Portugal

Parties	1987			1989			1994			1999		
Registered voters	7,813,103			8,107,694			8,490,367			8,600,643		
Voters	n.c.			n.c.			3,024,634	35.6		3,465,301	40.3	
Votes counted	5,490,302	70.3		4,016,756	49.5		2,928,073	96.8		3,351,675	96.7	
	V.C.	%	Seats	V.C.	%	Seats	V.C.	%	Seats	V.C.	%	Seats
PCTP-MRPF (extreme left/Communist-Maoist)	19,390	0.4	0	26,580	0.7	0	23,674	0.8	0	30,358	0.9	0
PXXI (extreme left/ecologists)							12,217	0.4	0			0
UDP (extreme left/Communist-Maoist)	52,613	1.0	0	44,907	1.1	0	18,807	0.6	0		BE	0
PSR (extreme left/Communist-Maoist)	28,824	0.5	0	31,621	0.8	0	17,720	0.6	0	62,022	1.9	0
POUS (extreme left/Communist-Maoist)				11,343	0.3	0				5,508	0.2	0
FER (extreme left)				7,644	0.2	0						
MUT (extreme left)	23,284	0.4	0									
PCR (extreme left/Communist)			0				2,585	0.1	0			
PEV (extreme left/ecologist)		CDU	0		CDU	1		CDU	0		CDU	
PCP (extreme left/Communist)	648,962	11.8	3	597,404	14.9	3	339,283	11.6	3	357,575	10.7	2
MDP (extreme left)	27,290	0.5	0	56,786	1.4	0	5,802	0.2	0			
PRD (left)	250,009	4.6	1			1						
PS (left/Socialist)	1,267,529	23.1	6	1,183,415	29.5	7	1,052,183	35.9	10	1,491,963	44.5	12
PPD-PSD (center/Liberal)	2,109,057	38.4	10	1,356,889	33.8	9	1,039,492	35.5	9	1,077,665	32.2	9

Table A2.13 (Continued)

Parties	1987			1989			1994			1999		
	V.C.	%	Seats	V.C.	%	Seats	V.C.	%	Seats	V.C.	%	Seats
CDS—CDS-PP (1993) (right/ Christian Democrat-Conservative-nationalist)	867,587	15.8	4	586,337	14.6	3	377,628	12.9	3	282,928	8.4	2
PDC (extreme right/ Fundamentalist)	40,459	0.7	0	29,718	0.7	0						
PDA (regionalist (Açores))							6,940	0.2	0	5,059	0.2	0
MPT (ecologist (diss. PPM))							12,442	0.4	0	13,669	0.4	0
PPM (miscellaneous/ monarchist)	155,298	2.8	0	84,112	2.1	0	8,236	0.3	0	16,164	0.5	0
PSN (miscellaneous/retired)							11,064	0.4	0	8,764	0.3	0
Total	5,490,302	100.0	24	4,016,756	100.0	24	2,928,073	100.0	25	3,351,675	100.0	25

Table A2.14 Spain

Parties	1987			1989			1994			1999		
Registered voters	28,43,7306			29,283,982			31,558,724			32,944,451		
Voters	n.c.			n.c.			18,664,053	59.1		21,209,685	64.4	
Votes counted	19,070,377	67.1		15,658,468	53.5		18,364,794	98.4		20,684,196	97.5	
	V.C.	%	Seats	V.C.	%	Seats	V.C.	%	Seats	V.C.	%	Seats
HB—EH (regionalist/Nationalist (Basque country))	360,952	1.9	1	269,089	1.7	1	180,324	1.0	0	306,508	1.5	1
PSA (left/regionalist (Asturias))												
AC (left/regionalist (Canary Islands))						0						
UPV (APV + PNPV) (left/regionalist (Valencia))						0						
UA (left/regionalist (Aragon))			0			0						
PSM (left/regionalist (Balearic Islands))			0			0						
EE (left/regionalist (Basque country))			0			1						
PSG-EE (left/regionalist (Galicia))		CIP	0		IP	0						
ENE (left/regionalist (Catalunia))	261,228	1.4	0	290,286	1.9	0						
TC-PNC (left/regionalist (Castille-Leon))							cf. PEP	cf. PEP		13,119	0.1	0
BNG (left/regionalist (Galicia))	53,116	0.3	0	46,052	0.3	0	139,221	0.8	0	347,205	1.7	1

Table A2.14 (Continued)

Parties	1987 V.C.	1987 %	1987 Seats	1989 V.C.	1989 %	1989 Seats	1994 V.C.	1994 %	1994 Seats	1999 V.C.	1999 %	1999 Seats
EU (right/regionalist (Extremadura))	39,369	0.2	0	cf. FPR			13,580	0.1	0	15,713	0.1	0
CA (PA) (left/regionalist (Andalusia))	185,550	1.0	0	295,047	1.9	1	140,445	0.8	0			1
PRC (center/regionalist (Cantabria))	14,553	0.1	0									
PRP (center/regionalist (La Rioja))				FPR		0						
UV (right/regionalist (Valencia))	162,128	0.9	0	151,834	1.0	0						0
PAR (right/regionalist (Aragon))	105,865	0.6	0							CE		0
CC (AIC) (center/regionalist (Canary Islands))	96,895	0.5	0			0			1	676,287	3.3	1
PANCAL (center/regionalist (Castille-León))	12,616	0.1	0			0						
UM (center/regionalist (Balearic Islands))	19,066	0.1	0						0			0
CG (PG) (center/regionalist (Galicia))	CUE		0	CN			CN					
EAJ-PNV (center/Christian Democrat-regionalist (Basque country))	226,570	1.2	0	303,038	1.9	1	518,532	2.8	1			1
EA (left/regionalist (Basque country) (diss. EAJ-PNV))			1			1			0			1
ERC (left/regionalist (Catalunia))	CEP		0	PEP		0	PEP		0	CN/EP		
PNG (left/regionalist (Galicia) (diss. CG))	326,911	1.7	0	238,909	1.5	0	239,339	1.3	0			0

Party	E1 Votes	E1 %	E1 S	E2 Votes	E2 %	E2 S	E3 Votes	E3 %	E3 S	E4 Votes	E4 %	E4 S
AV-MEC—ELS VERDS (EV) (ecologist/regionalist (Catalunia))				47,249	0.3	0	42,237	0.2	0	611,801	3.0	0
LVAM (ecologist/regionalist (Madrid))						0			0			0
AEG (ecologist/regionalist (Galicia))						0			0			0
PEE (ecologist/regionalist (Basque country))	CV	0.3	0			0			0			0
LOS VERDES-GRUPO VERDE (GV) (ecologist)				LISTA VERDE			VERDE			137,038	0.7	0
LOS VERDES (LV) (ecologist)	107,625	0.6	0	164,515	1.1	0	109,567	0.6	0	VERDES		0
CA (extreme left/regionalist (Aragon))						1			1			
IC (PSUC) ... (extreme left/post-Communist-regionalist (Catalunia))	IU			IU						299,851	1.4	1
PCE (extreme left/Communist)	IU			IU		1	IU		2	IU-EUIA		3
PASOC (left/Socialist)	1,011,750	5.3	1	961,742	6.1	1	2,497,671	13.6	7	1,213,254	5.9	1
PCC (extreme left/Communist-regionalist (Catalunia) (diss. PSUC))						0			1			0
PCPE (extreme left/Communist (diss. PCE))				79,918	0.5	0	29,692	0.2	0	25,304	0.1	0
PTEUC (extreme left/Communist (diss. PCE))	222,680	1.2	0	197,042	1.3	0						
PDNI (extreme left (diss. Izquierda Unida))										PROG		2
PSOE (+PSQ (left/Socialist)	7,522,706	39.4	28	6,276,554	40.1	27	5,719,707	31.1	22	7,420,035	35.9	22
CDS ... (center/Liberal)	1,976,093	10.4	7	1,133,429	7.2	5	183,418	1.0	0	41,496	0.2	0

Table A2.14 (Continued)

Parties	1987			1989			1994			1999		
	V.C.	%	Seats	V.C.	%	Seats	V.C.	%	Seats	V.C.	%	Seats
CDC (center/Liberal-regionalist (Catalunia))		CiU	2		CiU	1		CiU	2		CiU	2
UDC (center/Christian Democrat-regionalist (Catalunia))	853,603	4.5	1	666,602	4.3	1	865,913	4.7	1	934,259	4.5	1
PDP (right/Christian Democrat-Conservative)	170,866	0.9	0									
CP (AP)—PP (1989) (+UPN) (right/Conservative)	4,747,283	24.9	17	3,395,015	21.7	15	7,453,900	40.6	28	8,364,767	40.4	27
PAS—RUIZ-MATEOS (right)	116,721	0.6	0	608,560	3.9	2	82,410	0.4	0			
FN (extreme right)	122,799	0.6	0	60,667	0.4	0						
Other lists	288,758	1.5	0	472,920	3.0	0	148,838	0.8	0	277,559	1.3	0
Total	19,070,377	100.0	60	15,658,468	100.0	60	18,364,794	100.0	64	20,684,196	100.0	64

Table A2.15 Sweden

Parties	1995			1999		
Registered voters	**6,551,591**			**6,664,205**		
Voters	**2,727,317**	**41.6**		**2,588,514**	**38.8**	
Votes counted	**2,683,151**	**98.4**		**2,529,437**	**97.7**	
	V.C.	%	*Seats*	*V.C.*	%	*Seats*
Vänsterpartiet (VP) (extreme left/ post-Communist)	346,764	12.9	3	400,073	15.8	3
Miljöpartiet de Gröna (MP) (ecologist	462,092	17.2	4	239,946	9.5	2
Arbetarepartiet-Socialdemokraterna (SAP) (left/Social Democrat)	752,817	28.1	7	657,497	26.0	6
Centerpartiet (CP) (center/agrarians)	192,077	7.2	2	151,442	6.0	1
Kristdemokratiska Samhällspartiet (KDS) (right/Christian Democrat)	105,173	3.9	0	193,354	7.6	2
Folkpartiet Liberalerna (FP) (right/Liberal)	129,376	4.8	1	350,339	13.9	3
Moderata Samlingspartiet (MSP) (right/Conservative)	621,568	23.2	5	524,755	20.7	5
Ny Demokrati (NYD) (extreme right/populist)	2,841	0.1	0			
Sverigedemokraterna (SD) (extreme right/xenophobe)				8,568	0.3	0
Fairness list against the EU (miscellaneous/anti-Europeans)	14,904	0.6	0			
Free Criticism of the EU (miscellaneous/anti-Europeans)	18,398	0.7	0			
Sarajevo (miscellaneous)	26,875	1.0	0			
Other lists	10,266	0.4	0	3,463	0.1	0
Total	*2,683,151*	*100.0*	*22*	*2,529,437*	*100.0*	*22*

Table A3 Deputies in the European Parliament, October 1999 (breakdown by member state

	Germany		Austria		Belgium		Denmark		Spain		Finland		France	
GUE-NGL	PDS	6					SF	1	IU	4	VAS	1	PCF	4
													Indep	2
													LO	3
													LCR	2
42		6		0		0		1		4		1		11
V-ALE	GRÜNEN	7	GRÜNE	2	AGALEV	2			CE-PA	1	VIHR	2	Verts	9
					ECOLO	3			CNEP-EA	1				
					VU/ID21	2			CNEP-PNV	1				
									BNG	1				
48		7		2		7		0		4		2		9
PES	SPD	33	SPÖ	7	SP	2	SD	3	PSOE	22	SDP	3	PS	18
					PS	3			PDNI	2			PRG	2
													MDC	2
180		33		7		5		3		24		3		22
ELDR					VLD	3	V	5	CiU-CDC	2	KESK	4		
					PRL/FDF	2	RV	1	CE-CC	1	SFP	1		
51		0		0		5		6		3		5		0
EPP-DE	CDU	43	ÖVP	7	CVP	3	KF	1	PP	27	KOK	4	UDF	9
	CSU	10			PSC	1			CiU-UDC	1	SKL	1	DL	5
					CSP-EVP	1							RPR	6
					MCC	1							GE	1
233		53		7		6		1		28		5		21
UEN							DFP	1					RPF	12
30		0		0		0		1		0		0		12
EDD							FolkB	1					CPNT	6
							JuniB	3						
16		0		0		0		4		0		0		6
NI			FPÖ	5	VB	2			EH	1			FN	5
													RPF	1
26		0		5		2		0		1		0		6
626		99		21		25		16		64		16		87

GUE-NGL: Confederal group of the European United left/Nordic Green left
V-ALE: Greens/European Free Alliance
PES: Parliamentary group of the Party of European Socialists
ELDR: The European Liberal, Democrat and Reform Party

and political group)

Greece		Ireland		Italy		Luxembourg		Netherlands		Portugal		United Kingdom		Sweden	
KKE	3			RC	4			SP	1	PCP	2			VP	3
SYN	2			CI	2										
DIKKI	2														
7		**0**		**6**		**0**		**1**		**2**		**0**		**3**	
		GP	2	VERDI	2	GRENG	1	GL	4			GP	2	MP	2
												SNP	2		
												PC	2		
0		**2**		**2**		**1**		**4**		**0**		**6**		**2**	
PASOK	9	LAB	1	SDI	2	LSAP	2	PvdA	6	PS	12	LAB	29	SD	6
				PDS	15							SDLP	1		
9		**1**		**17**		**2**		**6**		**12**		**30**		**6**	
		Indep	1	LN	1	DP	1	VVD	6			LD	10	FP	3
				PRI	1			D'66	2					CP	1
				IDEMO	6										
0		**1**		**8**		**1**		**8**		**0**		**10**		**4**	
ND	9	FG	4	PPI	4	CSV	2	CDA	9	PSD	9	CP	36	MSP	5
		Indep	1	SVP	1							UUP	1	KDS	2
				FI	22										
				CCD	2										
				CDU	2										
				UDEUR	1										
				RI-DINI	1										
				PENSION	1										
9		**5**		**34**		**2**		**9**		**9**		**37**		**7**	
		FF	6	AN/SEGNI	9					CDS-PP	2				
0		**6**		**9**		**0**		**0**		**2**		**0**		**0**	
								GPV	1			UKIP	3		
								RPF	1						
								SGP	1						
0		**0**		**0**		**0**		**3**		**0**		**3**		**0**	
				BONINO	7							DUP	1		
				LN	3										
				MSFT	1										
0		**0**		**11**		**0**		**0**		**0**		**1**		**0**	
25		**15**		**87**		**6**		**31**		**25**		**87**		**22**	

EPP-DE: Group of the European People's Party (Christian Democrats) and European Democrats
UEN: Union for Europe of the Nations
EDD: Europe of Democracies and Diversities
NI: Technical group of independent members-mixed group. Nonattached.

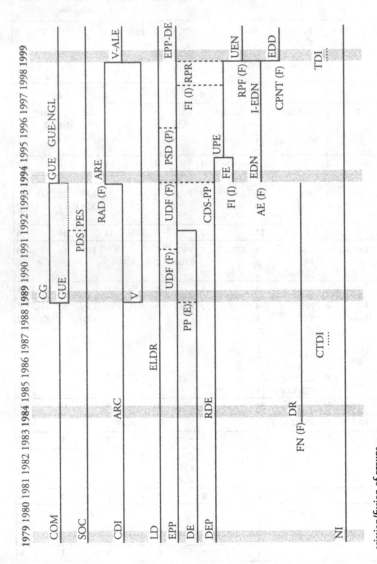

Chart A1 History of Political Groups in the European Parliament, 1979–1999

scission/fusion of groups ——————

principal movements - - - - - - -

Tables A4 (A4.1 to A4.6) Evolution of the political groupings in the European Parliament, 1979–1999

Table A4.1 European Parliament, 1979–1984 legislature*

Group	Total	Germany	Belgium	Denmark	France	Ireland	Italy	Luxembourg	Netherlands	United Kingdom
COM				SF 1	PCF 18		PCI 20			
					PCR 1		Indep 4			
	44	**0**	**0**	**1**	**19**	**0**	**24**	**0**	**0**	**0**
SOC		SPD 35	SP 3	SD 3	PS 20	LAB 4	PSI 9	LSAP-POSL 1	PvdA 9	LAB 17
			PS 4	SIUMUT 1	MRG 2		PSDI 4			SDLP 1
	113	**35**	**7**	**4**	**22**	**4**	**13**	**1**	**9**	**18**
LD		FDP 4	PVV 2	LV 3	UDF 17	Indep 1	PLI 3	DP-PD 2	VVD 4	
			PRL 2				PRI 2			
	40	**4**	**4**	**3**	**17**	**1**	**5**	**2**	**4**	**0**
PPE		CDU 34	CVP 7		UDF 7	FG 4	DC 29	CSV-PCS 3	CDA 10	
		CSU 8	PSC 3		CNIP 1		SVP 1			
	107	**42**	**10**	**0**	**8**	**4**	**30**	**3**	**10**	**0**
ED				KF 2						CP 60
				CD 1						UUP 1
	64	**0**	**0**	**3**	**0**	**0**	**0**	**0**	**0**	**61**
DEP				FRP 1	RPR 15	FF 5				SNP 1
	22	**0**	**0**	**1**	**15**	**5**	**0**	**0**	**0**	**1**
CDI			VU 1	FolkB 4		Indep 1	PR 3			
							DP 1			
							PDUP 1			
	11		**1**	**4**	**0**	**1**	**5**	**0**	**0**	**0**
NI			FDF 1				MSI-DN 4		D'66 2	DUP 1
			RW 1							
	9	**0**	**2**	**0**	**0**	**0**	**4**	**0**	**2**	**1**
	410	**81**	**24**	**16**	**81**	**15**	**81**	**6**	**25**	**81**

*Day of reference September 24, 1979.

Table A4.2 European Parliament, 1984–1989 Legislature*

	Total	Germany		Belgium		Denmark		France	
COM						SF	1	PCF	9
								PCR	1
	41		**0**		**0**		**1**		**10**
ARC		GRÜNEN	7	VU	2	FolkB	4		
				AGALEV	1				
				ECOLO	1				
	20		**7**		**4**		**4**		**0**
SOC		SPD	33	SP	2	SD	3	PS	20
				PS	5	SIUMUT	1		
	130		**33**		**7**		**4**		**20**
ELDR				PVV	2	LV	2	UDF	12
				PRL	3				
	31		**0**		**5**		**2**		**12**
EPP		CDU	34	CVP	4	CD	1	UDF	9
		CSU	7	PSC	2				
	110		**41**		**6**		**1**		**9**
ED						KF	4		
	50		**0**		**0**		**4**		**0**
RDE								RPR	15
								CNIP	2
								DCF	1
								PLI	1
								UDF-RAD	1
	29		**0**		**0**		**0**		**20**
GDE								FN	10
	16		**0**		**0**		**0**		**10**
NI				SP	2				
	7		**0**		**2**		**0**		**0**
	434		**81**		**24**		**16**		**81**

*Day of reference: September 10, 1984.

Greece		Ireland		Italy		Luxembourg		Netherlands		UK	
KKE	3			PCI	26						
KKE.es	1										
	4		**0**		**26**		**0**		**0**		**0**
				DP	1			GPA-PSP	1		
				UV/Psda	1			GPA-PPR	1		
				PCI/PDUP	1						
	0		**0**		**3**		**0**		**2**		**0**
PASOK	10			PSI	9	LSAP-POSL	2	PvdA	9	LAB	32
				PSDI	3					SDLP	1
	10		**0**		**12**		**2**		**9**		**33**
		Indep	1	PLI	2	DP-PD	1	VVD	5		
						PRI	3				
	0		**1**		**5**		**1**		**5**		**0**
ND	9	FG	6	DC	26	CSV-PCS	3	CDA	8		
				SVP	1						
	9		**6**		**27**		**3**		**8**		**0**
										CP	45
										UUP	1
	0		**0**		**0**		**0**		**0**		**46**
		FF	8							SNP	1
	0		**8**		**0**		**0**		**0**		**1**
EPEN	1			MSI-DN	5						
	1		**0**		**5**		**0**		**0**		**0**
				PR	3			SGP	1	DUP	1
	0		**0**		**3**		**0**		**1**		**1**
	24		**15**		**81**		**6**		**25**		**81**

Table A4.3 European Parliament, 1984–1989 legislature*

	Germany		Belgium		Denmark		Spain		France	
COM					SF	2	IU-PCE	1	PCF	9
							IU-PASOC	1	PCR	1
							IU-PSUC	1		
48	**0**		**0**		**2**		**3**		**10**	
ARC	GRÜNEN	7	VU	2	FolkB	4	CEP-EA	1		
			AGALEV	1						
			ECOLO	1						
20	**7**		**4**		**4**		**1**		**0**	
SOC	SPD	33	SP	3	SD	3	PSOE	28	PS	20
			PS	5						
165	**33**		**8**		**3**		**28**		**20**	
ELDR			PVV	2	LV	2	CiU-CDC	2	UDF	12
			PRL	3						
44	**0**		**5**		**2**		**2**		**12**	
EPP	CDU	34	CVP	4	CD	1	CiU-UDC	1	UDF	10
	CSU	7	PSC	2						
115	**41**		**6**		**1**		**1**		**10**	
ED					KF	4	CP	17		
66	**0**		**0**		**0**		**4**		**17**		**0**
RDE									RPR	15
									CNIP	2
									DCF	1
									UDF-RAD	1
29	**0**		**0**		**0**		**0**		**19**	
GDE									FN	10
17	**0**		**0**		**0**		**0**		**10**	
CTDI			SP	1			CDS	7		
12	**0**		**1**		**0**		**7**		**0**	
NI							HB	1		
2	**0**		**0**		**0**		**1**		**0**	
518	**81**		**24**		**16**		**60**		**81**	

*After elections in Spain and Portugal. Day of reference: September 26, 1987.

Greece		Ireland		Italy		Luxembourg		Netherlands		Portugal		United Kingdom	
KKE	3			PCI	26					PCP	3		0
KKE.es	1												
	4		0		26		0		0		3		0
				DP	1			GPA-PSP	1				
				UV/Psda	1			GPA-NCP	1				
	0		0		2		0		2		0		0
PASOK	10			PSI	9	LSAP-POSL	2	PvdA	9	PS	6	LAB	32
				PSDI	3					PRD	1	SDLP	1
	10		0		12		2		9		7		33
		Indep	1	PLI	3	DP-PD	1	VVD	5	PSD	10		
				PRI	3								
	0		1		6		1		5		10		0
ND	8	FG	6	DC	26	CSV-PCS	3	CDA	8	CDS	4		
				SVP	1								
	8		6		27		3		8		4		0
												CP	45
	0		0		0		0		0		0		45
Indep	1	FF	8									SNP	1
	1		8		0		0		0		0		1
EPEN	1			MSI-DN	5							UUP	1
	1		0		5		0		0		0		1
				PR	3			SGP	1				
	0		0		3		0		1		0		0
												DUP	1
	0		0		0		0		0		0		1
	24		15		81		6		25		24		8

Table A4.4 European Parliament, 1989–1994 legislature*

	Germany	Belgium	Denmark	Spain	France	Greece
CG					PCF 7	SYN-KKE 2 SYN-NAR 1
14	**0**	**0**	**0**	**0**	**7**	**3**
GUE			SF 1	IU-PCE 2 IU-PASOC 1 IU-PSUC 1	2	SYN-EAR 1
28	**0**	**0**	**1**	**4**	**0**	**1**
V	GRÜNEN 8	AGALEV 1 ECOLO 2		IP-EE 1	VERTS 8	
30	**8**	**3**	**0**	**1**	**8**	**0**
ARC		VU 1	FolkB 4	PA 1 CEP-EA 1	UPC 1	
13	**0**	**1**	**4**	**2**	**1**	**0**
SOC	SPD 31	SP 3 PS 5	SD 4	PSOE 27	PS 19 MRG 1 ADD 1 Indep 1	PASOK 9
180	**31**	**8**	**4**	**27**	**22**	**9**
ELDR	FDP 4	PVV 2 PRL 2	LV 3	CiU-CDC 1 CDS 5	UDF-U 12 UDF-C 1	
49	**4**	**4**	**3**	**6**	**13**	**0**
EPP	CDU 25 CSU 7	CVP 5 PSC 2	CD 2	PP 15 CiU-UDC 1	UDF-C 5 UDF-U 1	ND 10
121	**32**	**7**	**2**	**16**	**6**	**10**
DE			KF 2			
34	**0**	**0**	**2**	**0**	**0**	**0**
RDE					RPR-U 12 CNIP-U 1	DIANA 1
20	**0**	**0**	**0**	**0**	**13**	**1**
DR	REP 6	VB 1			FN 10	
17	**6**	**1**	**0**	**0**	**10**	**0**
NI				HB 1 CN-PNV 1 RUIZ-MATEOS 2	UDF-C app. 1	
12	**0**	**0**	**0**	**4**	**1**	**0**
518	**81**	**24**	**16**	**60**	**81**	**24**

*Day of reference: July 25, 1989.

Ireland	Italy	Luxembourg	Netherlands	Portugal	United Kingdom
WP 1				CDU-PCP 3	
1	**0**	**0**	**0**	**3**	**0**
	PCI 22				
0	**22**	**0**	**0**	**0**	**0**
	VERDI 3 / ARCOB 2 / DP 1 / La Drogua 1		REGEN-PPR 1 / REGEN-NCP 1	CDU-VERDES 1	
0	**7**	**0**	**2**	**1**	**0**
Indep 1	UV/Psda 1 / LL 2				SNP 1
1	**3**	**0**	**0**	**0**	**1**
LAB 1	PSI 12 / PSDI 2	LSAP-POSL 2	PvdA 8	PS 8	LAB 45 / SDLP 1
1	**14**	**2**	**8**	**8**	**46**
Indep 1 / PD 1	PL-PRI 3	DP-PD 1	VVD 3 / D'66 1	PSD 9	
2	**3**	**1**	**4**	**9**	**0**
FG 4	DC 26 / SVP 1	CSV-PCS 3	CDA 10	CDS 3	UUP 1
4	**27**	**3**	**10**	**3**	**1**
					CP 32
0	**0**	**0**	**0**	**0**	**32**
FF 6					
6	**0**	**0**	**0**	**0**	**0**
0	**0**	**0**	**0**	**0**	**0**
	PL PR 1 / MSI-DN 4		SGP 1		DUP 1
0	**5**	**0**	**1**	**0**	**1**
15	**81**	**6**	**25**	**24**	**81**

Table A4.5 European Parliament, 1994–1999 legislature*

		Germany		Belgium		Denmark		Spain		France	
GUE								IU-PCE	7	PCF	6
								IU-PASOC	1	Indep	1
								IU-PSUC	1		
	28		**0**		**0**		**0**		**9**		**7**
V		GRÜNEN	12	AGALEV	1	SF	1				
				ECOLO	1						
	23		**12**		**2**		**1**		**0**		**0**
PES		SPD	40	SP	3	SD	3	PSOE	22	PS	15
				PS	3						
	198		**40**		**6**		**3**		**22**		**15**
ARE				VU	1			CN-CC	1	RAD	13
	19		**0**		**1**		**0**		**1**		**13**
ELDR				VLD	3	LV	4	CiU-CDC	2	UDF-RAD	1
				PRL/FDF	3	RV	1				
	43		**0**		**6**		**5**		**2**		**1**
EPP		CDU	39	CVP	4	KF	3	PP	28	UDF	13
		CSU	8	PSC	2			CiU-UDC	1		
				CSP	1			CN-PNV	1		
	157		**47**		**7**		**3**		**30**		**13**
FE											
	27		**0**		**0**		**0**		**0**		**0**
RDE										RPR	14
	26		**0**		**0**		**0**		**0**		**14**
EDN						FolkB	2			AE	13
						JuniB	2				
	19		**0**		**0**		**4**		**0**		**13**
NI				VB	2					FN	11
				FN	1						
	27		**0**		**3**		**0**		**0**		**11**
	567		**99**		**25**		**16**		**64**		**87**

*Day of reference: August 1, 1994.

Greece		Ireland		Italy		Luxembourg		Netherlands		Portugal		United Kingdom	
KKE	2			RC	5					PCP	3		
SYN	2												
	4		**0**		**5**		**0**		**0**		**3**		**0**
		GA	2	VERDI	3	GRENG	1	GL	1				
				RETE	1								
	0		**2**		**4**		**1**		**1**		**0**		**0**
PASOK	10	LAB	1	PSI-AD	2	LSAP-POSL	2	PvdA	8	PS	10	LAB	62
				PDS	16							SDLP	1
	10		**1**		**18**		**2**		**8**		**10**		**63**
				RIF	2							SNP	2
	0		**0**		**2**		**0**		**0**		**0**		**2**
		Indep	1	PRI	1	DP-PD	1	VVD	6	PSD	8	LD	2
				LN	6			D'66	4				
	0		**1**		**7**		**1**		**10**		**8**		**2**
ND	9	FG	4	PPI	8	CSV-PCS	2	CDA	10	PSD	1	CP	18
				SEGNI	3							UUP	1
				SVP	1								
	9		**4**		**12**		**2**		**10**		**1**		**19**
				FI/CCD	27								
	0		**0**		**27**		**0**		**0**		**0**		**0**
POLAN	2	FF	7							CDS-PP	3		
	2		**7**		**0**		**0**		**0**		**3**		**0**
								GPV	1				
								SGP	1				
	0		**0**		**0**		**0**		**2**		**0**		**0**
				MSI-AN	11							DUP	1
				PSDI	1								
	0		**0**		**12**		**0**		**0**		**0**		**1**
	25		**15**		**87**		**6**		**31**		**25**		**87**

Table A4.6   European Parliament, 1994–1999 legislature*

	Germany		Austria		Belgium		Denmark		Spain		Finland		France	
GUE-NGL									IU-PCE	5	VAS	2	PCF	5
									IU-PASOC	1			Indep	2
									PDNI	2				
									PSUC	1				
	34		**0**		**0**		**0**		**0**		**9**		**2**		**7**
V	GRÜNEN	12	GRÜNE	1	AGALEV	1					VIHR	1		
					ECOLO	1								
	27		**12**		**1**		**2**		**0**		**0**		**1**		**0**
PES	SPD	40	SPÖ	6	SP	3	SD	4	PSOE	21	SDP	4	PS	15
					PS	3							Indep	1
	214		**40**		**6**		**6**		**4**		**21**		**4**		**16**
ARE					VU	1			CN-PAR	1			PRG	9
									CN-CG	1			Indep	3
									Indep	1				
	21		**0**		**0**		**1**		**0**		**3**		**0**		**12**
ELDR			LIF	1	VLD	3	LV	4	CiU-CDC	2	KESK	4	UDF-RAD	1
					PRL/FDF	3	RV	1			SFP	1		
	42		**0**		**1**		**6**		**5**		**2**		**5**		**1**
EPP	CDU	39	ÖVP	7	CVP	4	KF	3	PP	28	KOK	4	UDF	8
	CSU	8			PSC	1			CiU-UDC	1			DL	5
					CSP	1								
					MCC	1								
	201		**47**		**7**		**7**		**3**		**29**		**4**		**13**
UPE													RPR	17
	34		**0**		**0**		**0**		**0**		**0**		**0**		**17**
I-EDN							FolkB	2					RPF	8
							JuniB	2						
	15		**0**		**0**		**0**		**4**		**0**		**0**		**8**
NI			FPÖ	6	VB	2							FN	10
					FN	1							MN	2
													Indep	1
	38		**0**		**6**		**3**		**0**		**0**		**0**		**13**
	626		**99**		**21**		**25**		**16**		**64**		**16**		**87**

*After elections in Austria, Finland, and Sweden. Day of reference: May 15, 1999.

Greece		Ireland		Italy		Luxembourg		Netherlands		Portugal		Sweden		United Kingdom	
KKE	2			RC	3					PCP	3	VP	3	ALAB	1
SYN	2			CI	1										
				SV	1										
	4		0		5		0		0		3		3		1
		GP	2	VERDI	2			GL	1			MP	4	SSP	1
				RETE	1										
	0		2		3		0		1		0		4		1
PASOK	10	LAB	1	SDI	2	LSAP	2	PvdA	7	PS	10	SAP	7	LAB	60
				DS	17									SDLP	1
	10		1		19		2		7		10		7		61
				RIF	2	GaL	1							SNP	2
	0		0		2		1		0		0		0		2
		Indep	1	PRI	1	DP	1	VVD	6			FP	1	LD	3
				Indep	2			D'66	4			CP	2		
				PL	1										
	0		1		4		1		10		0		3		3
ND	9	FG	4	PPI	6	CSV	2	CDA	9	PSD	9	MSP	5	CP	15
				IDEMO	2									PECP	2
				SVP	1										
				UDR	3										
				FI	20										
				CCD	3										
				Indep	1										
	9		4		36		2		9		9		5		17
POLAN	2	FF	7	CCD	1			EVPN	2	CDS-PP	3				
				AN/SEGNI	1										
				LN	1										
	2		7		3		0		2		3		0		0
								GPV	1					UUP	1
								RPF	1						
	0		0		0		0		2		0		0		1
				AN	10									DUP	1
				MSFT	1										
				LN	4										
	0		0		15		0		0		0		0		1
	25		15		87		6		31		25		22		87

Maps (1 to 19) Political families in the 15 countries of the European Union (electoral results, June 1999)

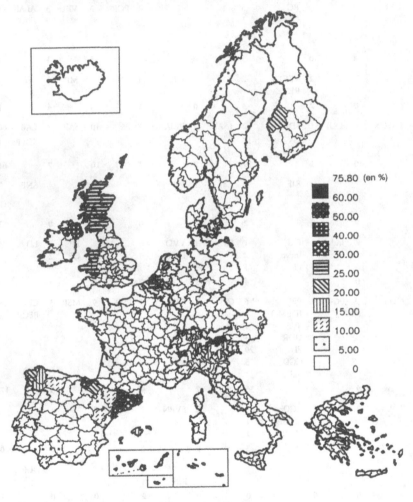

75.80 (en %)
60.00
50.00
40.00
30.00
25.00
20.00
15.00
10.00
5.00
0

Map 1 Electoral Results of Regionalist Parties

N.B.: *In the United Kingdom, the results are mapped on the scale of the major voting districts and not by county.*

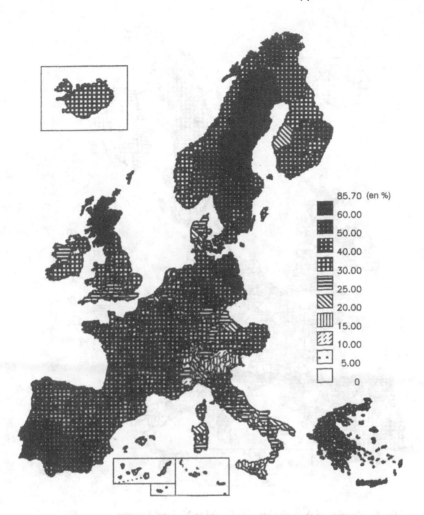

85.70 (en %)
60.00
50.00
40.00
30.00
25.00
20.00
15.00
10.00
5.00
0

Map 2 Electoral Results of the Left

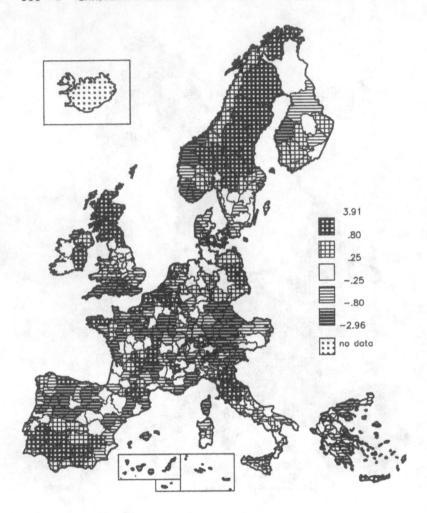

Map 3 Variation of the Left Vote with Regard to National Averages

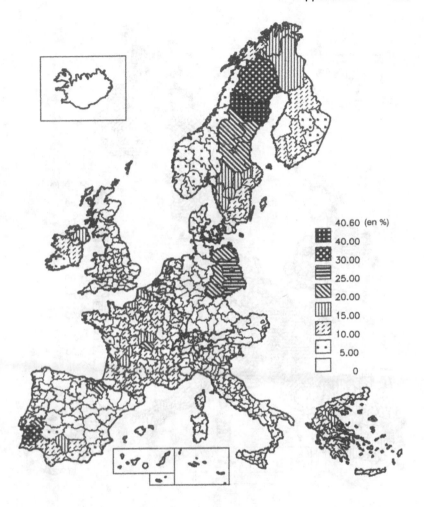

40.60 (en %)
40.00
30.00
25.00
20.00
15.00
10.00
5.00
0

Map 4 Electoral Results of the Extreme Left and Communists

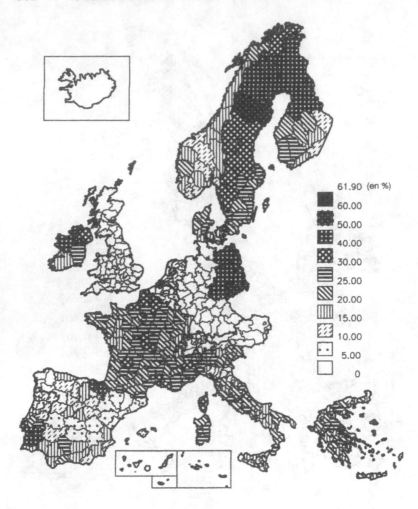

61.90 (en %)
60.00
50.00
40.00
30.00
25.00
20.00
15.00
10.00
5.00
0

Map 5 Proportion of the Extreme Left and Communists in the Total Left Vote

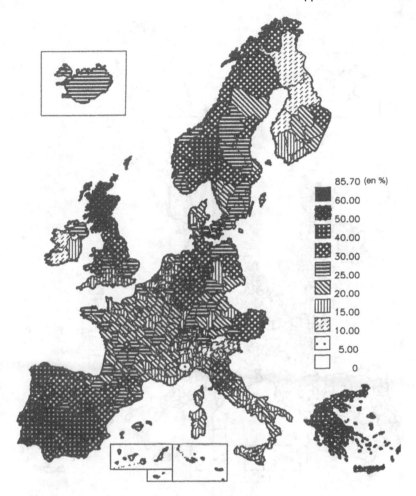

Map 6 Electoral Results of the Socialist and Labor Party Left

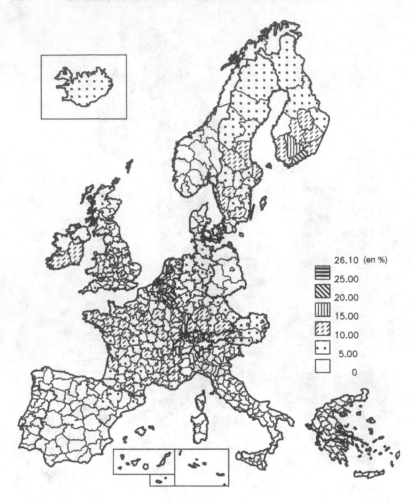

Map 7 Electoral Results of the Ecologists

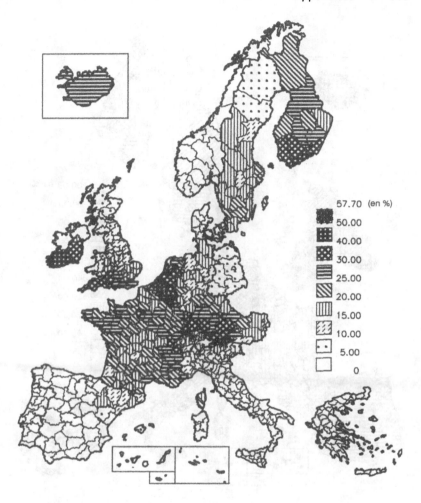

Map 8 Proportion of the Ecologists in the Total Left Vote

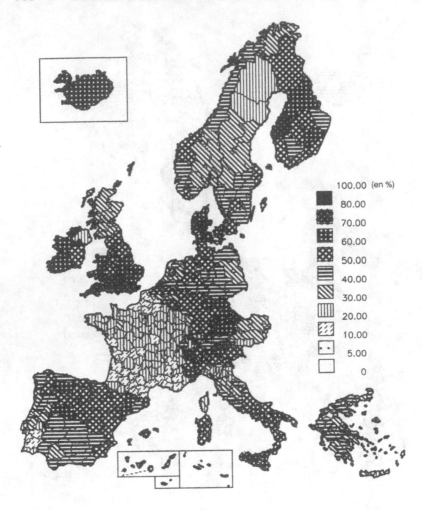

Map 9 Electoral Results of the Center Right and Traditional Right

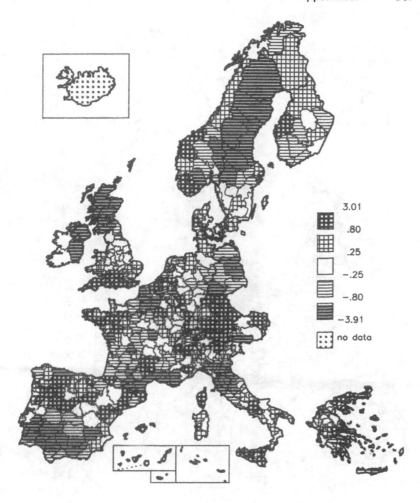

3.01
.80
.25
−.25
−.80
−3.91
no data

Map 10 Variation of the Center and Traditional Right Vote with Regard to National Averages

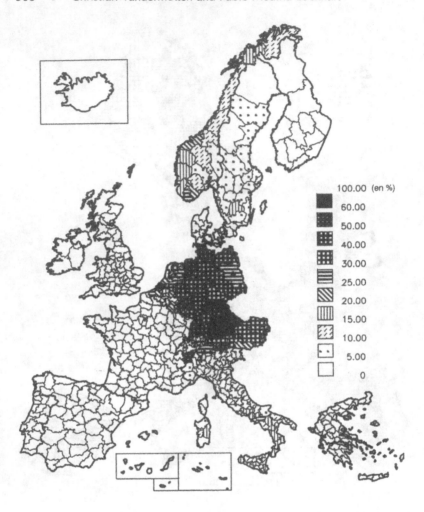

Map 11 Electoral Results of the Christian Democrats

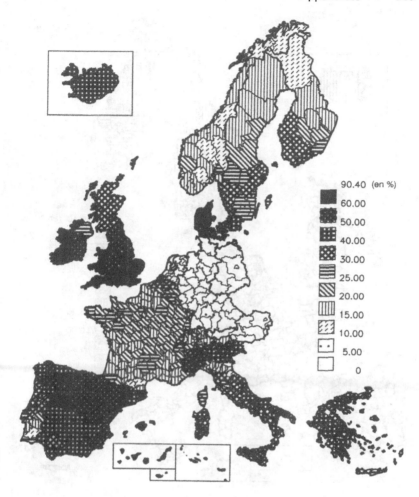

Map 12   Electoral Results of the Liberals and Conservatives

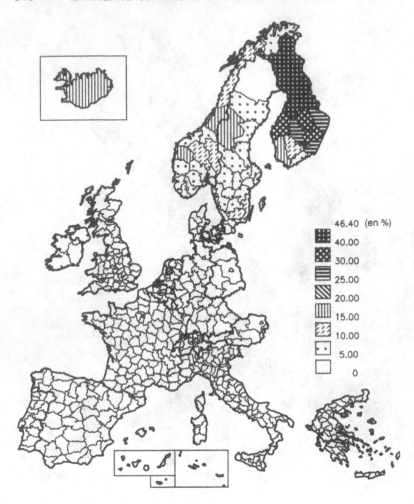

Map 13 Electoral Results of the Agrarians

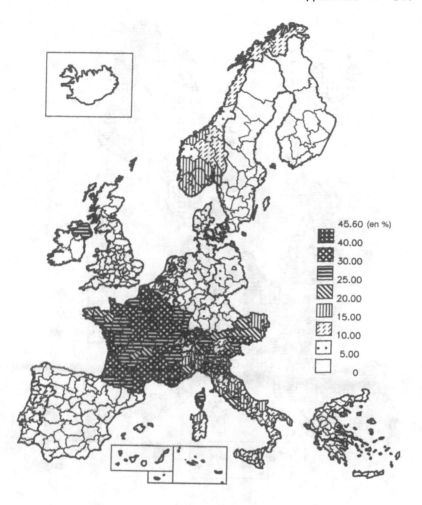

Map 14 Electoral Results of the Reactionary Right and Extreme Right

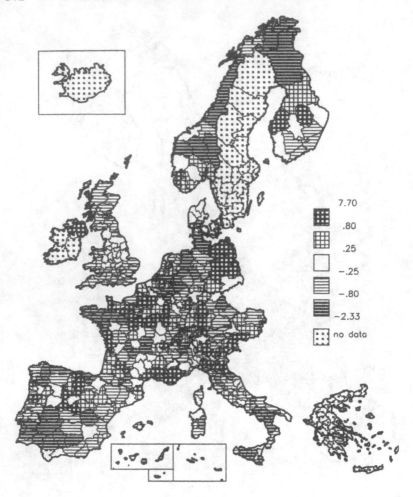

Map 15 Variation of the Reactionary Right and Extreme Right Vote with Regard to National Averages

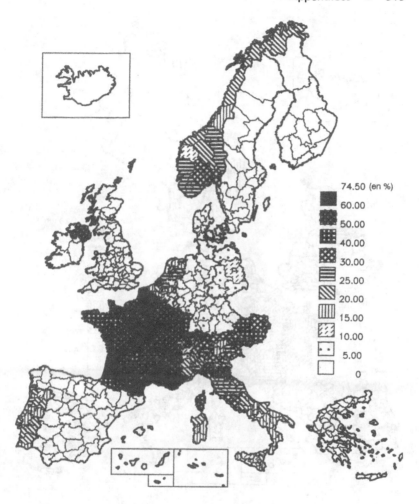

Map 16 Proportion of the Reactionary Right and Extreme Right Vote with Regard to the Total Center and Right

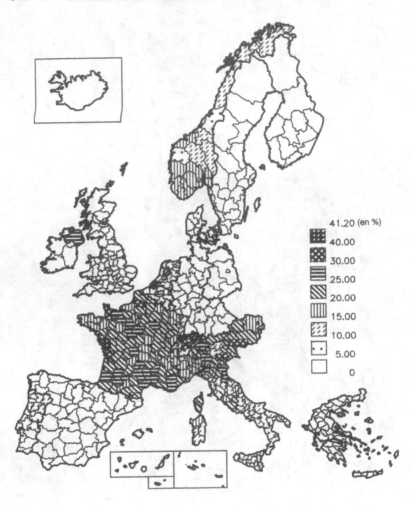

Map 17 Electoral Results of the Populists or Poujadist Reactionary Right and the Religious Extreme Right

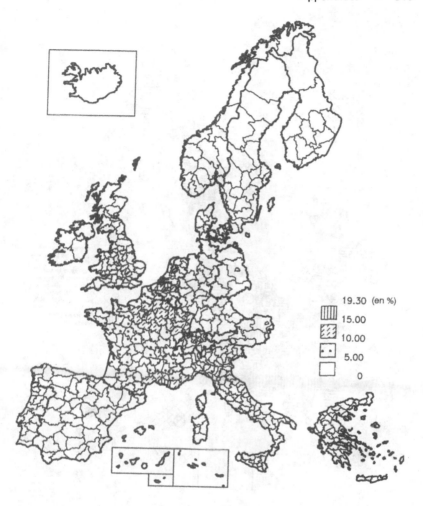

Map 18 Electoral Results of the Authoritarian Extreme Right

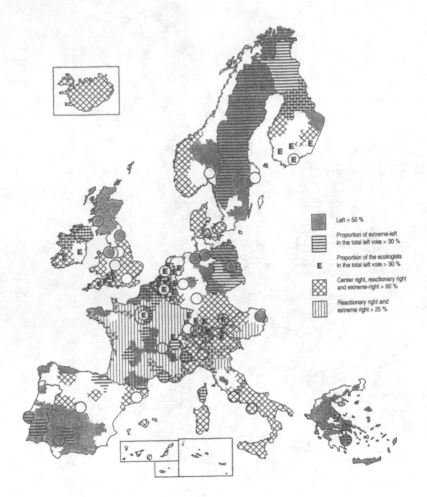

Legend:

Left > 50 %

Proportion of extreme-left
in the total left vote > 30 %

Proportion of the ecologists
in the total left vote > 30 %

E

Center right, reactionary right
and extreme-right > 60 %

Reactionary right and
extreme right > 25 %

Map 19 Synthesis of the Electoral Spatial Patterns in Western Europe

N.B.: *The circles indicate, where it is possible, the biggest towns and metropolitan areas.*

Index

318 • Index

Printed in the United States
By Bookmasters

Printed in the United States
By Bookmasters